Our Mysterious Universe

Mark Nelson

Contents

Universe As Consciousness 1

The Universe As Our Teacher 35

The Life Of The Individual As A Reflection Or Model Of Human Evolution 57

Where Have We Been (And Why Are We Still There) 78

Free Will Individualization 91

Evil 100

Curtain 107

Ufo And Devas 130

School Is Over 135

Looking Back From The Future 149

The Great Call 152

Universe As Consciousness

What is the sense of life? In general, does life have a meaning? And if so, what should we do with it? When we ask ourselves these three questions and look for answers to them, only then do we become human. I am writing these words, and outside the window is frost. I watch with delight how the millimeter-thick ice bas-relief gradually covers the lower two thirds of glass. After a minute or two, a picture appears that resembles lush summer vegetation: feathery leaves and intricately curved branches are clearly visible. Each "plant" is unique and at the same time perfectly inscribed in the composition: it does not leave empty spaces and does not obscure the neighbors. A perfect picture is not the fruit of a perfect plan?

It is impossible not to think about the deep meaning of what I see: the artist who created this amazing work - "ordinary" frozen water (an inorganic substance that has neither genes nor DNA). What kind of energy lies behind such phenomena and what kind of consciousness does it need to have in order to plan and create such beauty? Few people will be able to draw such a perfect pattern on their own, and it will take much more time than a couple of minutes. I am even more surprised that the existing belief systems, shared by supposedly the most advanced nations on the planet, not only cannot convincingly explain most of the mysteries of nature (this is just understandable), but usually prefer to ignore them and even try to deny many phenomena that do not fit into framework of their ideologies. Ignoring and denying is perhaps the best thing that people can count on who dare to draw the attention of others to such realities, The main trouble of our traditional religions and science is not limited knowledge and not even an overestimation of the degree of one's own understanding of reality. The biggest mistake they make is when they attack those who are able to perceive a much widerand a perfect universe and who are trying to cooperate with this universe in spreading the Light, thereby expanding human knowledge far beyond the bounds of rigid belief systems. What prevents us from saying to ourselves: yes, we stilldon't know much? Why is it bad to admit that there is a mysterious world around us? In addition, it is known that the cosmological systems of orthodox science and orthodox Western religions largely contradict each other and even, in essence, exclude each other. (We will soon back.)

And yet, I want to express my position from the very beginning: I believe that both science and religion are right in something important - they simply view reality from different positions. But if science, with all its rationality, lacks wisdom, and religion, with all its wisdom, does not stand up to reasonable analysis, then they do not agree with the highest, universal Truth. After all, the Universe in which we live (and we can see it for ourselves around us) is reasonable, expedient, wise and, most importantly, loving. This is what I will try to show. Here are just a few examples of anomalous phenomena that have a deep (and therefore disturbing) meaning and are therefore dismissed by our establishment as unworthy of serious study.

There are many cases when the physical body of a

person was separated from higher "bodies" - in a state of clinical death, under the influence of drugs or high speeds (when falling, in a centrifuge), in a state of shock, etc. People who were considered unconscious by others, observed their physical body from the side and could subsequently accurately describe the events that took place. We all have dreams, and sometimes visions of a different kind, that tell a lot about our internal states (undetected diseases, complexes, etc.), or about what we can expect from the future, and also tell us how to behave further (if we are not too lazy to analyze them). There are a lot of reports about the so-called poltergeist, possession and other parapsychic phenomena. Over the past half century (in fact, throughout history), all over the world, people who can be trusted have seen UFOs. And many - at one level or another - had contacts with "aliens".

In different countries of the world, the so-called "crop circles" spontaneously appear - huge pictograms of the most diverse and beautiful geometric shapes. Everyone can see them, and not all cases turn out to be fake.
Throughout human history, there have been spontaneous spontaneous combustion of people, and all attempts to reproduce this phenomenon artificially have failed. Memories from past lives, appearing in many people, may indicate the repetition of life. Sometimes children give such details about people and events in the past, or about distant places that they could not possibly have known.

This list can be continued for a long time. Many books have been written and many photographs and videos have been made documenting these so-called "anomalous" phenomena. But instead of honestly examining them and expanding our knowledge of this amazing universe, the establishment shows a complete reluctance to listen to anything that might disturb them thoroughly organized belief systems (although the latter are clearly imperfect and are increasingly being proven wrong). Fortunately, now - as happens periodically on any planet - new, fresh energies are coming to our Earth, and people from various walks of life are beginning to be skeptical of the old explanations, realizing in their hearts that there is much more to life than our public institutions.

So, let's repeat what has been said: the cosmological systems of orthodox science and orthodox Western religions contradict each other in many ways and even, in essence, exclude each other. One system is based on the erroneous belief that the physical plane and its associated phenomena are all that really exists. (And everything that exists happened by chance!) Another system, common in several religions, essentially claims that everything was created by some capricious and very cruel deity for no clear reason (the qualities and desires attributed to this god always strangely correspond to the ideology of the ruling circles). The powers that be tend to try to be one foot in every camp, and it's very important for them to deny, ignore, and refute everything that science and religion can't explain.

It's the nature of human systems beliefs, our ideologies, our establishment - they pretend to have all the answers to attract and retain adherents and thereby perpetuate its existence by "maintaining order". And we ourselves, small personalities, are still very immature, and we like to believe that we are much smarter than we really are. To think that we, or any other person, or any human belief system, has all the answers is not a sign of ignorance? Conversely, the first sign of wisdom is the understanding that we still have a lot to learn. But, since we are still at a relatively early stage of human evolution, it often happens that "The blind lead the blind." What is left for a normal thinking person to do if our cultural paradigm is designed for schizophrenics? (Actually, this is more of a Siamese twin paradigm, because many people are comfortable with both belief systems at the same time.)

Given the above, people can be divided into two categories: some are always ready to perceive new aspects of the Truth that are constantly being revealed to humanity. Others hold on to "good old" beliefs and resist anything that undermines them, not realizing that, historically, these beliefs are relatively recent. I would call the first group "thinkers", and the second - "believers". It can be assumed that agnostics and atheists who are proud of what they have "scientific" or "skeptical" view of reality, fall into the category of thinkers, not believers. But it is not always the case. We are constantly faced with the fact that the scientific establishment is as stubbornly defending its dogmas and opposing anything unorthodox as any fundamentalist religion. And that's the whole point. Obviously, in order to expand your knowledge of life, you must at least allow the possibility of restructuring your own worldview when new truths (scientific or religious) are discovered, and not automatically reject what is incomprehensible to us.

Let's start with religion. When you seriously study the essence of many great religious beliefs - deeply and without prejudice - it becomes clear that there is much more in common than disagreement. Disagreements and discrepancies appear after the inspired teacher is gone. After all, if there is a "God", then is it possible to imagine that a Being worthy of that name will reveal the whole truth for all time only once - to the chosen people in one place - and ignore all the rest? If there is a God, then we are all His children, and He loves us equally. If there is a God, then He, like the sun, shines on everyone. Therefore, a wise person constantly evaluates the "tradition", using his insight and intuition to understand the difference between true enduring wisdom that contributes to the spiritual evolution of mankind, and what over time has become just another meaningless dogma that does not help future enlightenment in any way. So, maybe the whole kaleidoscope of worldviews on our planet, including new revelations that come in continuously, are pieces of a giant puzzle? What if you don't build an impenetrable wall around every little fragment, rejecting everything else, as many belief systems do, How about a glance from the top of a mountain? Shall we not then see that each fragment emphasizes some particular aspect of the universal truth?

Now about orthodox science. If you do not believe in God, can you believe that ordinary human scientists can know everything? Many believe that the current scientific theories of evolution have already explained life on Earth in detail from the very beginning to the incredibly complex current state. But don't many scientific truths, born only a century ago, look somewhat primitive and even absurd today? Don't we now realize that decades will pass and many of today's scientific truths will look just as stupid? Also keep in mind that scientific theoriesbegin with axioms and postulates - that is, initial positions that are not self-evident, but are accepted without proof. Take any materialistic theory and follow its logical chain: in the end you will come across an unconfirmed basis, and everything will end with one miracle being interpreted by other miracles.

Surprisingly, many scientists believe that science already knows quite well how the universe was formed and how the universe functions, and it remains only to clarify the details. But this is far from true! However, this very conviction indicates that soon profound new (for us) truths will be given to mankind. Because this is how the universe enlightens us. First, some truth is revealed. Then, when it finally becomes accepted and "orthodox", another truth is revealed, which replaces the old one. This happens endlessly, and it always leads to the expansion of human consciousness. We are given an idea, it is deposited in the human mind and gradually becomes a universally recognized ideal, which eventually crystallizes into an ideology. By that time, the time is already approaching for the introduction of a broader idea into humanity. This process is repeated again and again, and as a result, humanity gradually becomes more and more enlightened.

Let no one think that this book is against science! I want to make it clear from the very beginning: it is scientists who in the near future will scientifically confirm the presence of dimensions of being outside the physical world. Finally, everyone admits that people do indeed have many psychic abilities.faculties now denied by materialistic science. It is extremely important to realize that at higher levels "Spiritual Science" has always existed! It is this precipitation of available knowledge in human consciousness over long periods of time that has always supported the continuous growth of human intelligence and wisdom, which in turn has fueled our evolution. As we continue to absorb the higher truths, we will continue to move further and further away from the animal stage and move even faster towards a higher consciousness - towards enlightenment, predicted by the teachers of mankind.

I am fully convinced that deep truth can be found at the core of all great religions. And undoubtedly, scientists have already made countless discoveries and will continue to do so. These discoveriesled and will lead to a significant increase in human knowledge. Acting together, these two branches of human research (science and religion) can and must make, and will certainly make, the most important contribution to the enlightenment of mankind. Enlightenment of mankind will come when we realize our potential of Intelligence, Wisdom, Love. Eternal wisdom expanding through constant insights, will lead to an even better understanding of universal truth and free us from the burden of ignorance.

The universal truth is what I would like to talk about in this book. This is the truth that reflects the absolute reality of our universe. The truth that all serious researchers are trying to discover. Truth that embodies self-evident signs truth: consistency, consistency, consistency. The truth, which, although eternal, continues to be revealed as the consciousness of mankind grows. And most importantly: this is the Truth that resonates with our highest, deepest, sacred essence - with our Heart, with our Soul. This is her main characteristic.

The reason for writing this book was nothing less than a desire to helpbringing a much-needed new cosmological paradigm to life! This new paradigm is now taking hold all over the planet. We all have a choice: we can take advantage of this new tremendous opportunity to expand our Consciousness (Life) and become an important part of these new energies. Or we can continue to live in relative ignorance, choosing what suits us from the limited belief systems of our culture and letting others think for us. And once again we ask: What is the meaning of life? In general, does life have a meaning? And if so, what should we do with it? These three questions are actually three aspects of the Unified Search.

That is what we are looking for. And if you take part in this most important of activities, you will never look at the world the same way. In the following pages, I have tried to bring together some of the deepest and most essential knowledge that is available to Man. Knowledge gained from the best teachers and from the best teachings of the past and present, confirmed (and expanded) by life experience. In a word, this is the kind of knowledge that leads to Wisdom. The acquisition of such a quality as Wisdom, along with Love, is the main goal of the human wave of life in which we are now.
This book should find a response in your Soul, in your

Heart. Since this is so, it cannot contradict the Higher Mind, because the Soul and the Higher Mind are united in the human Being. Everything in this book that does not resonate in your Heart, in your Soul, in your intuition, discard it! Accept only what resonates with your Higher and Best Self.

But I must say right off the bat: there is nothing really new in this book. Concepts that may seem unfamiliar to many peoplealways existed in a teaching known by many names: Eternal Wisdom, Ancient Wisdom, Esoteric Teaching, etc. When those in power tried to suppress this knowledge, it was preserved thanks to secret societies. In addition, many of its elements can be found in the world's scriptures, especially when they are read at the Soul level!
The divine teachers of mankind have always emphasized: the more a person becomes enlightened, the deeper meaning is revealed to him in their teachings. Therefore, as our consciousness grows, we begin to see not only the literal meaning of the scriptures. These sermons and stories corresponded to the intellectual level of the average person living at the time they were written down. But there were also higher truths in them, waiting for people to wake up and see their meaning.

Much of what we will talk about can also be found in the books of the great thinkers and philosophers of all times. And some ideas, perhaps in the form of insights, visited you yourself. And, of course, I would not want all this to be accepted by anyone as a new gospel. In no case! And without that, there is no shortage of people who are trying to convince you that the belief system in which they happened to believe is the only one, and that only in it can you find answers to all questions.

(And the more they subconsciously doubt this, the more they work to convince others, and along with themselves.) The last thing you need (and you won't find in this book) is more guidance on what to believe. This is just a presentation of my understanding of reality - no doubt limited and imperfect. In general, I advise everyone who has reached that level of development of consciousness at which people begin to read such books, to approach any text critically and without prejudice. (We'll talk about the importance of the critical hike later.)

So, in this book you will find a comprehensive (albeit briefly stated) "worldview" (moreover, the "view" of both the external and internal world), which you can compare with any other worldview, and most importantly - with your own life experience. Even if at this point in your life you are convinced that life has no purpose, keep reading. We will talk about the fact that this stage also fits into the great meaning of Life. What if we humans didn't just believe what we're told, but tested reality through our own experience and observation, sometimes accepting conventional wisdom and sometimes looking for better explanations?

What if all claims about the meaning of life are wrong and we need to learn to see the answers for ourselves? What big Truths will we receive from small truths, when - a little later in this book, we will discuss the following issues, very different and sometimes quite mundane: If the cells of our body are very often updated, then why is it already in middle age begin to show signs of aging? Why do we get old at all? Why is death good for the human race, and why should we not try to eliminate

natural death? (Let's assume it's within our power.)
Why in the embryonic state do humans (and other animals) repeat the earlier stages of animal development?

Why do babies have wrinkles (and fingerprints) on their hands even before birth?

Why is gender ambiguity sometimes found among people? (And why is it more common now than before?)

Why do some people devote their lives to altruistic service, while others become greedy tyrants (strong and yet petty)?

Why any normal person can usually tell the difference "false" note, even without a musical education, and why are there "false" notes? Why is there a direct relationship between music, sound, mathematics and even organic growth?

Why is it said that creative, insightful people have "taste"? Why is sport needed and why is it so popular? How is it that almost everywhere right under the surface of the planet there is clean drinking water?

Why are minerals - metals, minerals, coal, oil, etc. - most often found in the form of "deposits" scattered around each other? from a friend over long distances?

If this is not enough for you, do not despair: perhaps we will talk about many other issues that interested you. And in the process of discussing them, this book will show that the universe is not just "friendly" to us: it our true friend. Yes, our Universe is a benevolent, patient, wise in everything, loving Being. A being who takes to heart our highest and best thoughts. Maybe I'm reading your mind. You think: how can you say such a thing! History remembers so many bloody events! Yeah "friendly" universe!

Yes, we have all experienced pain and loss, some less, some more. But painful as the human phase of our long journey may be, if we see the broader picture of cosmic evolution, we will realize that our (relative and temporary) suffering has its causes, as well as our joys. All this is a necessary part in our conscious evolution and the evolution of our merciful Universe. It may be hard to believe, but we all play a role in "Divine Plan", or in the "Great Comprehensive Plan", as it is also called. The world that is given to us is incredibly beautiful and amazing.

And, most importantly, we must recognize that most of our (human) problems are our own creation. This means that the only way to rise higher and not cause more pain to ourselves is to raise consciousness. The growth of consciousness is one and often the only solution of all problems!

And again (for the last time): What is the sense of life? In general, does life have a meaning? And if so, what should we do with it? Every conscious person seeks to know this. Every person needs to know this! To know:

We must first understand that we will always be a part,

a growing part of this wonderful - incredible - absolute blessing called Life.

Life is an ever-expanding state in which you have always been and always will be (whether in the physical body or out of it).

Life experienced as the Eternal Now.

Life allows and encourages, in fact even demands that we realize our potential and fulfill our destiny. Our destiny involves the constant growth of consciousness so that we can become no less than co-creators, along with all other living forms within the greater Life!

Life much more important and much more complex than we can imagine. And, most importantly, our great Life will lead humanity to a wonderful future that is open to us and awaits only our balanced decision and action!

Life this is Everything: what we so often, without thinking and not appreciating, take for granted. We must understand and awaken to the realization that the little life we experience is a gift, coupled with the duty of absolute Life, which embraces the entire known and unknown universe, all that exists, the Cosmos. Some call it God.

In setting our priorities, however, we have significantly deviated from talking about new energies that have an impact on our planet. Let's get back to this new one.

Approximately every two millennia, a new layer of

teachings is introduced into the consciousness of mankind, and gradually most people become supporters of the new paradigm. These higher truths come from the higher Realms and from the higher Beings who govern the human race. Here is one of the main concepts of the current new paradigm: we do not live in a universe of matter and space, but, in essence, in a universe of energies. Remember: there is no such thing as dense "matter"!

What we take for matter is only the result of the activity of energy at the lowest and grossest level. And although science has recently recognized this important truth, only a few of the most enlightened scientists (and their number is growing) realize that energies have a quality that could be called consciousness. Let's put it differently: energy is the result of the activity of consciousness. What we perceive as matter is, in fact, energy (consciousness) at the lowest level.

What is a level? Let's talk about this in more detail, because this issue is also very important. Everyone knows that we exist and express ourselves on different levels. We have a physical body and we express ourselves physically; we have emotions and we express ourselves emotionally; we have a mind, and therefore we are able to think rationally. But many of us do not understand that our emotional and mental bodies are just as real as the physical body, and that they exist on their levels (planes, spheres) in the same way that our physical body exists on the physical plane. And, although they are usually associated with our physical body in the waking state, they can exist without it.
It is understood that these are the spheres (bodies) in

which "we" inhabit during sleep (and also after the death of the physical body). But the corresponding aspect of us lives in these fields (spheres) even when we are awake. In the waking state, these fields (spheres, bodies) go a little beyond the limits of our physical body and can be perceived from the outside as our "aura".

All our energy bodies (both lower and higher, spiritual) together form our energy field, our true "I".
Orthodox-minded scientists are trying to prove that there is only a physical plane and that all our various emotions and thoughts are born of physical causes. They will never prove this: chemical elements, like other matter, are not able to think and feel the way we do on a human level. What is true is that these finer energy bodies penetrate deeply into our the "physical" body when we are alive and awake.

Our physical body itself is only a lower and grosser form of energy. To see this, consider cases where people are severely injured and "pass out" (permanently or temporarily), even if the brain was not physically damaged. Conversely, there are cases when a person has a serious brain injury or even has a significant part of the brain removed, but the mental ability is notdecreases and he still retains thinking skills. Doesn't this indicate that we have a mind that does not depend on the brain for its existence, but that uses the brain as a means of functioning in the physical world?

Much remains to be learned about so-called "mental retardation" in the future. I don't think that in most cases the personality or the mind is retarded; rather,

this mental body does not agree with the physical body enough, perhaps due to physical injuries. Or it may be because the Higher Self, or Soul, is pursuing its own goals. One possible reason for "mental retardation" could be that over many lifetimes the mind has become too dominant and has actually blocked the love aspect.
In such situations it may beit is desirable to "put aside" the mind (to some extent) for a period of a lifetime, so that the energy of Love (Heart) can flow freely and bring more harmony to a living being.

It is quite obvious that the real threats to humanity come from those whose heart, or "body of love", is defective! Not from thosewho has deficiencies in the mental, emotional or physical body. We need to understand that our physical world and our physical sensations are just a (relatively) low and gross form of energy, and in fact they are like a distorted shadow of the higher worlds. And, most importantly, we must develop a higher consciousness in ourselves in order to understand these higher worlds. Only then will it become much easier to comprehend other realms of reality. This is especially true of the spiritual planes or worlds. Yes, there are huge, higher (some call them spiritual) planes, or worlds (or spheres? dimensions? fields?), and the inner world of the individual vaguely and at a much lower level reflects them.

Now let's be clear about what we mean by "spiritual planes or worlds." Apart from all the associations we may have with the word "spiritual", it primarily refers to specific levels of consciousness that are related to, but transcend to, the realms of consciousness in which we normally inhabit. In other words, in whatever dimension (world) a certain being lives (mineral, vegetable, animal, human, in the world of the Soul, etc.), beings in higher kingdoms in a certain sense perform a "spiritual" evolutionary function in relation to to beings that are in the realms of lower levels. This means that we humans can be considered "spiritual" in relation to the lower realms.

Therefore, becoming moreenlightened, we will begin to bear greater responsibility for them. In the same way, those who are above us on the wave of life (we call them guardian angels or guide spirits, the Spiritual Hierarchy, etc.) are responsible for helping us in our evolution. When our consciousness grows, when we become wiseand loving beings and will be initiated into the next higher realm (the realm of pure Love-Wisdom), we will no longer perceive it as a spiritual heaven, but simply as our usual habitat. (We'll talk about this later.)

Let's look at it from a different angle: if some great Divine Being (whose normal habitat is the spiritual world) descended to a lower level, which, however, for us still remains spiritual, then for this great Being it would be a tragedy, downgrading, if you will. World scriptures and myths tell us that this really happened (although rarely). Of course, we are not talking here about those who sacrifice themselves by incarnating in the human kingdom in order to help our further enlightenment. We emphasize once again: speaking of "spiritual levels", we simply mean higher levels of consciousness in which we do not yet consciously dwell and which therefore we cannot fully understand. Of course, these spiritual realms do not in the least resemble a naive children's picture in which beautiful people sit on clouds andlistening to the music of harps, and angels watching over them flutter around.

All the teachers and inspired writings tell us that this higher spectrum of Life is perceived as brighter and more significant than the realms we now inhabit. And, although we will find that life in these higher realms brings much more joy, our spiritual search will continue there. When someone deserves the right to enter (or move to) this plane of existence (and it will eventually happen to all of us through our efforts over many lives), he is convinced that this is the level of the very best human qualities - and much more. It is the seat of the abstract mind—the highest correspondence of the discriminating mind—where intuitive understanding (its sometimes called direct knowledge).

This is the Kingdom where wise Love and loving Wisdom reign supreme! Compassion, altruism and pure reason fill the atmosphere. This is "Heaven", where everyone is united by a fiery, focused, purposeful Will to serve the Divine Plan. These are the three main aspects, or the three Divine Energy Beams. Space! In those rare moments when we reach our highest state of joyful loving consciousness, when we experience our subtlest thoughts, we only touch the lower reflection of this true home of our spiritual Self (we will talk about this later). But it should be noted that those beings who have surpassed the physical level in their development and whose consciousness is concentrated in these, as we call them, Spiritual worlds, perceive everything in a completely different way, not in the same way as we do. Of course, this is to be expected, because their perspective is much higher and wider than ours.

Another important point: all you, I, or anyone else really knows is our thoughts and feelings. Ultimately, it is impossible to prove with absolute certainty that anything exists other than consciousness. You don't have to think long to be convinced of this. But "mind games" are not the intention of this book. There are many important reasons why what we perceive as the outside world exists, and this should be taken seriously. Let's get back to energy.

As we begin to realize that "everything is energy," that all energy has the potential to be good or bad (for us), and that whatever we come into contact with affects us in some way, we begin to see the differences much better between forces. Any place, any person, tree, weather, noise, song, color - everything, to a certain extent, either contributes to the growth of our consciousness, or slows it down. So, when someone begins to realize that Everything is Energy, and to learn the language of energy is the most important step in the spiritual evolution of this personality! We can understand energy as what we perceive on the level of the physical senses, but the really significant energies are extremely subtle and can only be felt with the help of our higher (spiritual) energy bodies (and their centers) that have the appropriate vibration frequencies. A small digression.

The above explains why we should, whenever possible, use the "gifts of nature" in their natural state - when the energies are best balanced and complement each other, resulting in the most beneficial effect. We must understand that the whole is by no means the sum of its parts! The whole, and the whole alone, contains the whole inner essence of Life. That is why when we take a natural product apart and try to isolate, concentrate and collect its essence, often a lot is irretrievably lost. Such stupidity has already done us a lot of harm: diseases, drug addiction, other addictions, etc. Whether it is "physical" or "subtle" energies, whether we are trying to isolate vitamins from food or light energy from sunlight, we must understand: We must understand that even the lower forms of energy are not just blind forces: they have their own rhythm of vibration and they correspond to the higher manifestations of energy.

For example, it is known that the proportions in our solar system (the orbits of the planets, etc.) are directly related to what we perceive as musical harmony, geometric shapes, mathematical ratios, and so on. It is due to the omnipresence of correct proportions and ratios that people subconsciously perceive some sounds and forms as beautiful, and others as "ugly", and ultimately learn to usecorrect proportions and relationships in all their affairs. This alone should be enough to show the biggest skeptics that the whole universe is based on a single idea, a plan. Let's clarify: the Divine Plan. If we talk about creation, it is known that in various religious traditions everything begins with a word or sound. The sound initiates or, by at least accompanies the onset of physical manifestation. It's right. Sound, audible or inaudible, accompanies the creation (and destruction) of matter, just as light (and still higher energieselectromagnetic range) is a creator at the highest levels. When this vibration that accompanies the universe reaches full harmony, we will have a symphony of spheres, the cosmos will come to full completion, and we will be able to immerse ourselves in silent peace.

To summarize: Matter-Space = Energy = Consciousness; it's all the same, but it's perceived differently at different levels of enlightenment. However, Consciousness is still primary; in fact, this is the universe. Everything is Conscious Life! Yes, every atom, molecule and cell, every stone, every plant, not to mention every galaxy, star or planet - everything is endowed with its ownits own inherent energy, its own form of consciousness. In addition, what we call "space" actually symbolizes the highest level of Consciousness. It is said, "God dwells in the gaps." If so, what significance does this have for science (or "art") of astrology?

If we lived in a universe of matter, then the principlesastrology would be difficult to recognize in any way reliable. On the other hand, if the whole universe consists of conscious energies (in fact, of great Beings) that form a cosmic unity, this, of course, is self-evident.by itself does not yet prove the basic principles of astrology, but at least offers a context in which the energies of what we perceive as cosmic bodies can affect us and our planet. If gravity, sunlight and the "solar wind" known to us, cosmic rays and many other known and unknown forces affect our planet at lower levels (these influences can be measured withwith the help of currently existing, still imperfect, instruments), can not stellar or planetary energies also have an effect on us at higher levels that is not yet measurable by instruments? Our young humanity has not even begun to study the myriad energies and forces that form our cosmos. There are other levels and ranges of being that we cannot yet even imagine.

Let's see where this line of reasoning leads us. If (as the Teachings of Wisdom state) the Universe is the infinite expanse of Life, the Cosmic Mind that encompasses all levels of consciousness and extends from the "dreamless sleep" of stone to the incomprehensible, grandiose fiery mind of the great "Lord" of the galaxy - and beyond that So, what exactly is consciousness? Of course, this is something much more and vastly different from anything that we humans can comprehend with our very limited minds today. The impossibility of determining the qualities of that consciousness possessed by higher, lower or parallel kingdoms is obvious: for this we need to have a comparable level of consciousness. Since humanity occupies only a tiny part in a very large range of Consciousness-Life, there is no need to talk about it.

At the first attempt to give a definition of consciousness, we will immediately encounter the severe limitations of our European languages - languages primarily of commerce and technology, almost alien to the Spirit. The meaning attached to our word "consciousness" is reduced to the realm of reason and feeling, because it is here that humanity polarizes, and therefore the word itself cannot mean anything that goes beyond these functions. But language shapes (and limits) our concepts!

In addition, people involved in physics are usually focused in their concrete (lower) mind and perceive everything at this level. They are not able to see clearly at the higher, abstract levels of human consciousness, and therefore it is difficult for them to comprehend these more subtle worlds. (There are reasons for this, and we'll talk about it later.) As soon as our consciousness expands and rises to such a level that it already captures the sphere of love-wisdom (a very important sphere!), we begin to understand what a huge potential we have and what huge higher gifts await us.

We may not immediately understand this, but when we begin to relate to life with a sense of responsibility and goodwill, we enter the Path (which we ourselves create) - the highest spiritual path that everyone is talking aboutreligion. A responsibility. Good will. Attentiveness. Thanks to them, wisdom is gradually acquired over many lifetimes. With effort and over time becoming wise and pure enough, we eventually stop being self-praise animals and begin to experience and live our inner Divinity. In this way, we acquire both the desire and the ability to become true servants of the planet.

At this most important step, we begin to fulfill our destined role in the human kingdom, that is, we become conscious co-creators! And together with other beings from all kingdoms, with spiritual support, we begin to work on the process of implementing the Divine Plan. We know how this has happened throughout history through the biographies of extraordinary personalities - those artists, philosophers, spiritual teachers and scientists who helped and are helping to develop our true civilization. These highly developed beings are often called luminaries or torches, because they have an inner Light that reflects a high degree of wisdom and pure intelligence, unattainable for most people. But you need to know that it is in this direction that the bulk of humanity is now gradually rushing, and this process will continue in the coming era. It is interesting to note that many of these people probably did not even know that they were helping planetary evolution.

We may think that consciousness is the accumulation of what we have absorbed through our senses and processed with our minds. But I repeat: the highest enlightenment comes to us through our higher centers, energy centers, which in some traditions are called chakras (we will talk about this later), and not through our physical senses. Since our planet is surrounded and permeated by countless energies emanating from cosmic and solar sources, as well as from the thought forms of our planetary lives on all levels, the analogy with tuning a radio receiver will be appropriate: we choose which of these waves to "catch". But we also radiate ourselves! That's why it's so important to take care relate to our thoughts. After all, the mind is the "builder" on a mental level, and we must be careful about what we build. And that's why sincere, selfless prayer and meditation can tune us into higher vibrations (rhythms), thereby helping us to "absorb the Light."

Let's take a closer look at the light analogy as applied to the level of spiritual growth. Light in the literal and figurative sense of the word begins with maximum freedom. Coming into contact with matter (impregnating matter, if you like), he loses some freedom, but at the same time raises the "consciousness" of matter. The penetration of Spirit into matter creates consciousness. Then, over time, these spiritual energies separate that part of matter that has received the Light, thereby allowing it to ascend, or continue its growth, in the kingdom where it was - mineral, vegetable, animal, human or other. The remaining unenlightened part is left to wait for the next wave, and this process continues until finally everything is "liberated", or reaches "perfection".

This is the true evolution, the evolution of consciousness. Liberation of matter! Modern scientific theories claim that the universe "slows down" (the second law of thermodynamics), but in fact it is just the opposite: the lower consciousness (what we perceive as matter) rises to the higher (spiritual) consciousness. "Matter" turns into energy — Spiritual Energy. The real Universe comes to life more and more. And we are part of it all! We may also think that "matter" exists only on the physical plane, but the realms of consciousness also have their own grosser or lower levels. So something analogous to the process described above takes place in all dimensions as the enlightenment work "One Life" overcomes the inertia of these lower, coarser energies.

Another important secret: a characteristic feature of all energy in Our Conscious Universe is the desire for balance and harmony. This is one of the ways of the Cosmos to the final perfection. And on the physical plane, this is carried out thanks to the well-known law of action and reaction. We must understand that, like all physical laws, it has higher correspondences on higher planes. In the human kingdom, balance and harmony are ultimately achieved through justice. This means that nothing "passes without a trace" - by our actions we either multiply what is given to us, or take away from these gifts. Ultimately, everything balances out. Indeed, "what we sow, then we shall reap!"

On the levels occupied by our personalities (physical, emotional, mental), the manifestation of this law in time is called karma. We are earning and will continue to earn either "positive" or "negative karma" depending on our actions. It is important to understand that karma exists not to punish us, but to teach us. And when we reach a level where we use our mind, love and wisdom so as not to provoke wrong actions (reasons), we will notwe will have to suffer more from the counteractions (consequences) of the forces that we set in motion. Let us now ask ourselves the question: can we even try to comprehend the infinite, these higher realms, the Mind of God? Of course we can't!

But we can discern some details of the Divine aspects and attributes on our lower level of existence. This brings us back to the source of Everything: Cosmic Life, where everything "lives and moves and has its being" (see Acts 17:28). How can we, who are only at the human stage of the Divine path, know the Unknowable? What can we know about the absolute Deity of all religions, about the Universal Principle and the "Laws of Nature," as scientists call it,about this Living, All-Wise, All-Loving, Infinite Universe in which we and everything else have such an important role to play? Primarily: Trying to find out something about universal (that is, universal) energies, we again and again collide with the numbers "three" and "seven", with trinity and septenary.
Here are some examples of the seven in the universe:

The seven colors of the rainbow.
Seven notes.

Seven types of crystal structures.

"Seven holes" in the human head.

Seven main energy centers-chakras.

Seven age periods of life (we will talk about this later).

Seven wonders of the world.

Seven days of creation and seven days in a week. Even the seven deadly sins.

And this list can go on and on. As for the trinity: from a scientific point of view, every energy, everything manifested consists of polarity and the force generated by this polarity. The positive and negative poles and the force generated by them are always triplicity, starting with the atom and up to the Cosmos as a whole. Another quality that every expression of Life has is that in everything, including the entire Universe, activity and apparent calm alternate. In the Teachings of Wisdom, this is called manifestation (manifestation) and pralaya, respectively. In the near future, scientists will learn much more about the universality of this phenomenon.

In religious teachings around the world, the numbers three and seven are very common. Everywhere it is said that the Absolute Unity, or God, manifests itself in three aspects. In our own human kingdom, we can understand these three aspects as:

1. Divine Will;

2. Divine Love;

3. Divine Mind.

All religions are based on this Trinity and deify it in the form of personified Deities. In patriarchal Christianity, this is the Father, Son and Holy Spirit, in orthodox Hinduism - Shiva, Vishnu and Brahma, in other religions - the divine Father, Mother and Child, etc. They are connected with the first three Cosmic Rays. At the higher levels, four additional qualities (or Rays) are attributed to the Third Ray, the Divine Mind. Taken together, they make up seven. Let's name additional Rays:

Ray 4: Harmony-Beauty by effort or struggle; Ray 5: Concrete Knowledge;

Ray 6: Idealism and devotion;

Ray 7: Organization and creative Ritual or Rhythm. In other words, higher, spiritual consciousness:

7) perfectly organized,

6) represents an ideal in any situations

5) has everything knowledge

4) creates perfect beauty and harmony,

3) deeply intelligently and actively expresses himself,

2) wise, benevolent, full of love,

1) has the Will and Power to ensure that everythingit was possible.

These signs correspond to the Seven Divine Rays. The seven rays can be divided into three rays of aspect and four rays of attributes. These seven conscious energies, which permeate the entire Universe and, among other things, determine the qualities of our personalities, come from one immutable, unknowable Principle - let's call it that for lack of a better word. Many religions of the world call him God.

Later in this book we will continue to speak of the three major cosmic rays of energy, as well as an additional four, which together make up the spiritual septenary. Remember the "seven spirits before the throne" (see Rev. 4:5)? Three and seven - these numbers are found again and again in both religious and secular teachings. It is very important to know that all life in the universe - from stone to the solar system - arises under the influence of these seven most powerful Rays of cosmic energy, acting in one combination or another.

In other words, in our Conscious Universe, the Seven Rays are the driving force behind evolution. They give the necessary impetus for all life to develop further, to its next step. There are no good or bad rays. Any energy can be misused! The result depends on many factors. If we talk about how this manifests itself in a person, then the main factor is the level of spiritual consciousness achieved. For example: The person of the "First Ray" the one who demonstrates the Ray of Will and Power is full of the energy of these qualities. At one pole it can be a tyrant who dominates through strength, control, cruelty and values only power over others. On a higher turn of the evolutionary spiral, the people of the First Ray, being leaders by nature, use their will to help humanity and move it forward. This may be through leadership in politics or other areas, allowing them to mobilize others for difficult but necessary tasks.

The "Second Ray" person demonstrates the qualities of Love-Wisdom and can be either a weak, fearful or harmlessly sentimental person, or one who exemplifies compassion, altruism, courage and wise insight in helping humanity. These are the qualities of the Heart. A person charged with the energies of the "Third Ray" of Reason and Activity can scatter energy on meaningless deeds or try to manipulate others for his own benefit. But if he is an enlightened person to a certain extent, then he uses his mental abilities to best coordinateenergy to raise the level of human civilization. This beam is associated with "Law of Economy" (which manifests itself as efficiency).

The people of the "Fourth Ray", the Ray of Harmony through Beauty (or Conflict), are not boring, they love to argue and can even be quarrelsome. They like to take risks, they quickly get bored with security. But they are creative people, often dramatic and flamboyant, who can create incredible beauty in form, music, literature, drama, etc. (It is not uncommon for actors and other creative people to have a quarrelsome nature.)

But the man of the "Fifth Ray", on the contrary, can sometimes seem boring. Because it is the Ray of Concrete Knowledge or Science. In the worst case, such a person can get bogged down in insignificant trifles. But this Ray (like the fourth) is the Ray of the human kingdom. It is he who leads us to become thinking beings. This Ray guides humanity towards technology and information (and away from the focus on emotions and desires). Now such influence is very necessary.

The man of the "Sixth Ray" can lead us into the abyss of the narrow-minded fanaticism - or, if this is an enlightened person, to the heights of the greatest ideals. After all, this is the Ray of Idealism and Devotion. It has had a strong influence on mankind over the past few centuries.

And finally, the Seventh Ray is the Ray of Organization and Ritual. He is now beginning to influence our entire planet and has already given us (among other things) the type of bureaucrat who sees nothing beyond his rules and regulations. But thanks to this same Ray, both large and small groups and organizations will arise that will give people the opportunity to realize their potential. And, what is very important, the energy of the Seventh Ray will allow humanity to know and use the rhythms and rituals of Life!

We have all met people who fit the above descriptions. But most often people demonstrate the qualities of more than one ray. The fact is that our physical body, and the emotional (astral) and mental bodies, and the lower "I" (personality), and the Soul itself have their own ray. Their combination determines what we will be in incarnation. And it is very important to highlight their subtle essence from our aforementioned aspects! Knowledge of the Seven Rays began to be revealed to the human mind at the end of the nineteenth century. Perhaps this is the main and most important sacrament of those that are manifesting outside today.

Much information is now available on the Seven Rays, and it will be very helpful to become familiar with it. If, comprehending Divine energies and delving into new revelations that are now available to human consciousness, you experience shock and fear, remember the "bright" (or enlightened) side of the coin. Think of the glorious future that humanity has in store if we do not miss this opportunity to raise and further expand our consciousness. Of course, some will prefer to remain "attached" to their old ideologies and belief systems and will not take advantage of new energies and new opportunities for change and growth. But let's think about it: do we want to remain "cavemen"? They, too, were probably content with their primitive beliefs. So, here are the most important points that I wanted to cover in the first section:

The Universe (Cosmos) as a whole is a conscious energy. The Universe (Cosmos) as a whole is Unity. This Unity is manifested in the Universe as seven Cosmic Rays of energy. The Universe (Cosmos) strives for balance and harmony, which manifests itself in the human kingdom as justice. All Life is endlessly replacing each other states of activity and outer peace.

We will explore these and other topics in more detail

later in the book. But first, we must clarify something for ourselves, without which our upward progress is impossible.

The Universe As Our Teacher

Somewhere in the lab, a cute white mouse is running nimbly through the maze. This little rodent knows his way and knows what awaits him at the end - he has already happened to be there more than once. Quite confidently and without any problems, he gets where he wants. Almost without stopping, he rises on his hind legs, presses a small button with his small nose and watches in pleasant anticipation how grains of food fall from somewhere above. If we could read mouse thoughts, then perhaps we would now know how proud this animal is that he has learned to get tasty, satisfying food. At the same time, he has no idea about people (they are outside his field).vision) who are now watching him and who conceived and staged this experiment.

Let's think: are we humans so different from this mouse? We live our lives, "discover" our discoveries, "invent" our own inventions (and get our own food). Don't we take credit for our results? At the same time, we do not know the truth that there are much wiser and more developed beings who are watching us from other dimensions. Higher beings who come up with ideas that promote our progress and come up with new learning situations that will take us - individually and collectively - to the next stage of our evolution. Many inventors and researchers admit that they have been helped by "flashes" of intuition, dreams, or insights. It is also known that many inventions and discoveries were made simultaneously in different parts of the earth by people who (consciously) did not contact each other.

We have come to our second main theme: the universe

that we humans perceive with our mind and five physical senses is nothing but a perfectly organized learning environment. Yes, what appears to us as an endless expanse of space with occasional inclusions of cosmic matter ("macrocosm"), as well as our own physical bodies ("microcosm") is actually a teacher. The teacher is so perfect, wise and loving that, through whatever realm of nature a "unit of consciousness" evolves (mineral, vegetable, animal, human or other) and at whatever level of development this unit is, its surroundingthe environment will certainly be used by its Higher Self to lift this individual to the next level of enlightenment. Every event, every experience that we have in life provides us with an opportunity to learn something. Very often the experience is repeated over and over until we finally learn from it.

And again, let's talk about the need to develop awareness. The theater of life is not only events ("play"), but also a stage with scenery, which is also necessary for the play to take place. The life of the mineral, vegetable, and animal kingdoms teaches us as much as the heavens. But the most important thing, as already mentioned, is to develop the quality of discrimination throughout life. Discrimination contributes to the perception (and ultimately to the creation) of the correct proportions and relationships in all things. On the physical plane, proportion and right relationships give what we perceive as true beauty, and beauty is one of the lowest manifestations of Cosmic Love. Take, for example, art (any): true art arises due to the fact thatthe artist applies discrimination in choosing and combining the right proportions and ratios, the result of which is beauty. And beauty is just one of the ways the universe teaches us the importance of these qualities: distinction, proportion, consistency. Real art in all its forms, from architecture to weaving, is the lowest form of cosmic Love created by man (on the physical plane). Therefore, our creations are the highest manifestation of a purely physical form. We have all heard that the sculptor, when working with a stone, cuts off everything unnecessary in order to release the beauty contained in it. Maybe this applies to all manifestations of love: it is everywhere, only it needs to be released? Perhaps it is the same in music: the composer does not use all possible sounds at once, but chooses from their variety only beautiful and, The bottom line is this: we need to release the encoded Spiritual Love and allow it to strengthen our own rudimentary Love. We must remember: what we perceive as "goodness, truth and beauty" in our lower world is nothing but the lower reflection of Reason, Wisdom and Love in the spiritual world!

And, of course, developing in ourselves the ability to distinguish between the correct ratios and proportions, we must learn to discard everything that does not contribute to "goodness, truth and beauty." We see the process taking place: in the lower realms (including our own body), what is useful is absorbed and the rest is rejected. And what is "not useful" in the higher kingdoms may be very good for the lower ones (a kind of closed food chain). This is how what we call "the grace of nature" develops. On a higher astral plane (emotions and desires), one of the ways to manifest Love is the art of correct human relationships. On the mental level, one of the ways to manifest Love is the art of higher mathematics.

Let us repeat once again: any genuine art, no matter what sphere it belongs to, is a lower reflection, or lower correspondence, of the higher spiritual reality of pure Cosmic Love. It requires a distinction that leads to proportionality and right proportions. Thus, when we become aware of the Universe as a teacher, one of the first and most important insights that are suggested to us is correspondences, or similarities of relationships.

Here are some examples of correspondences: awakening and sleep correspond with life and death; seasons - with periods of life; the life of an individual is comparable to the evolution of humanity as a whole. (We'll talk more about this shortly.) As a matter of fact, everything that we in our physical existence perceive as "good, true and beautiful" has a higher correspondence - some important spiritual reality! This is nothing but a universal law - the Law of Correspondence: "As above, so below." Since there are correspondences within all levels of consciousness on which we are, and between them, it is precisely "above" that is the Reality, and "below" (the physical world with which we identify) is a virtual reality, more like a shadow!

We will continue throughout this book to give examples of correspondences that indicate that Life is a medium of endless potential lessons. Speaking about the fact that the universe is our teacher, let's not forget about one more great help given to humanity: about those great enlightened Beings who, of their own free will, bring enormousa sacrifice to promote evolution on our planet and in particular in our human kingdom. But before we talk more about these great Souls, Let us first emphasize that there are ultimately only two philosophical approaches to the problem of absolute reality.

a) The materialist school maintains that the universe has no apparent purpose. Everything that exists, including human thought and feeling, is made of physical matter-energy - or is a consequence of its work. AND, as far as we know at present, earthly humanity is the highest form of intelligence in the universe.

b) According to the spiritual approach, the universe has a purpose. In addition to the physical dimension of reality, there are others. These worlds are inhabited by Beings (or Lives) with other levels of consciousness that can (and do) influence humanity.

c)
There is a widespread belief among spiritualists that at least some of these Beings (who live in higher dimensions, or higher planes) are much wiser and have much greater abilities than humans. Many also believe that at least a few of these Beings voluntarily joined together in a group (something like a spiritual planetary ashram). And these Divine Beings have taken it upon themselves to provide moral assistance to humanity, not interfering with our free will, but facilitating movement in the direction that is consistent with the Divine purpose of the Universe. In various religious traditions of the world, members of this group are called differently: saints, angels, teachers, etc.

Since they are beyond our concepts of gender and form,

we will simply refer to these enlightened Elders as Spirit Guides or the Spiritual Hierarchy of the planet. (And one of the goals of this book is to help, albeit a little, but to inspire others to help, these Divine Beings in Their efforts to lead humanity to the realization of its cosmic destiny.) It is also very important to realize that we receive Divine guidance not only from other Beings; we also have, and have always had, our own Inner Guide, our Higher Self, who wants to help us make the most of our opportunities.

In different traditions and belief systems, there are different names for this aspect of our big "I": superconsciousness, transpersonal "I", Soul, Solar Angel, Guardian Angel, etc. In this book they will be used as synonyms. But it should be emphasized that we humans have an individual Soul, while the subgroups of the lower kingdoms (animals, plants, minerals) have a soul "group". (Watch the behavior of flocks of birds, schools of fish, swarms of insects, etc., and you will understand a lot about it.)

But back to people. As soon as we begin to understand that we have our own personal higher guidance, to live in harmony with this great Being and receive instructions from him (in fact, the entire Universe we perceive is the physical expression of the Great Being), tremendous changes begin within us. We begin to perceive events and objects from the point of view of their internal energy, and not their external manifestation, and we try to understand what lessons we should learn from all this. Of course, not only obvious "messages" from the Universe, but also the most subtle ones can teach us a lot. For example, our

Soul often creates situations in space and time that we perceive as coincidences, but in fact they are planned. We must always be sensitive to suchevents (scientifically called synchronistic)! This is one of the most common ways to guide and help us in life. Much has been written about synchronicities. You can probably remember their examples in your own life. At some point, you experienced a pleasant (or unpleasant) surprise. It was only much later, in retrospect, that you understood how this event contributed to your personal growth. It is difficult to overestimate the importance of the right timing - both when we plan and when we evaluate the events of our lives.

Knowledge of the ongoing processes leads a person further and further into the world of wisdom, and this is precisely the world - the spiritual world. With the accumulation and use of wisdom, the speed of our evolution increases dramatically! Here's what it means: By becoming wise enough to begin to tap into these ever-present opportunities, we progress much faster in our spiritual enlightenment and experience the pangs of ignorance much less frequently. Moreover, when this very an important aspect of enlightenment, life becomes much clearer and we begin to live and act in a state of greater peace, harmony, efficiency and with ever-increasing self-control, if you will. As already mentioned, this is the most important step in our evolution, as a result of which there is a clear acceleration.

Speaking of "evolution": we keep repeating this word, but what actually evolves? Orthodox science believes that it is a physical form that gradually improves and adapts to its environment. There is some truth in this, but in fact, the consciousness betrayed to us that lives inside us, our true "I", is evolving. In the evolution of the physical form (even in individual life) we observe only corresponding changes. I remember many years ago I heard this phrase: "When you are over forty, you have the face you deserve." I think there is something in that too. It's not that a person with finer facial features is necessarily more developed spiritually, because there are many other factors involved. But in general, when a person becomes more enlightened, this is reflected in appearance.

The physical form of man on earth was gradually changing; this process is likely to continue. But the most significant changes occurred in mental abilities: at the service of our ever-expanding consciousness was an ever larger and more complex brain. Anthropological data show that each new type of person was marked by a less robust physique, but was more sensitive. Some might argue that as athletes continue to set new strength and endurance records, we humans are actually getting stronger. But new records are set due to the fact thattechnique improves, skills are honed, and only for a short time in the physical flowering of an athlete, and not at all because all of humanity is becoming stronger. Not even the strongest man can last five seconds in a duel with a gorilla of the same size, not to mention large predators.

If "survival of the fittest" (physically) is the driving force behind evolution, then why have we humans lost virtually all body hair - even those living in the coldestarctic regions? One can hardly speak of physical adaptation here. But if the driving force is the expansion of consciousness, then this loss makes sense. Primitive man was simply forced to use his primitive mind in order to learn how to survive through the ability to build a dwelling and make clothes for himself, and most importantly, to tame fire. If you like, we were forced to "wiggle our brains", and this act every time helps us expand our consciousness and, ultimately, become more spiritually enlightened.

Wiping out the entire human kingdom would be relatively easy, but try to get rid of all the flies or cockroaches! It is generally accepted that a bacterium, an earthworm or a daisy is much more adapted to life than we, more complex creatures. So let's not talk about natural selection anymore. Any thinking person who looks at the past (or present) with open eyes will see many examples where circumstances have inspired or even forced us humans to expand our intelligence. We will continue to become more knowledgeable and wiser, and more capable of love. Ultimately, life has one goal: Enlightenment. And all our experience serves this purpose! Let's talk more about the evolution of consciousness.

Like everything else in the universe, our physical planet is designed to continuously lead us to the next stages of enlightenment. Most people take both the physical structure of the Earth, and the apparent randomness of the location of forests, seas, the distribution of minerals in the bowels, etc., as a matter of course. But behind this imaginary accident lies a higher goal. Note that during that period of time in human history, when we finally reached the initial stage of mentality, we immediately "discovered" metals and deposits of coal and oil; learned how to turn the sap of certain trees into rubber and produce transparent solids (glass). This list goes on. Wasn't it inevitable (with a little help from above) that people soon learned how to make machines and vehicles? All this is not as prosaic as it might seem at first glance. But due to the fact that we acquire knowledge unconsciously and because "the closer you know, the less you respect", we perceive the mostamazing circumstances as something ordinary. And absolutely in vain. Many wise people have pointed out that sometimes the smallest details determine whether life on the planet, as we understand it, can exist. And if so, in what direction will it evolve.

Here are some examples. In order for coal to form (the fuel without which the industrial revolution is unthinkable), the vegetable kingdom had to evolve (that is, in fact, grow in terms of consciousness) to the stage of trees. Then it was necessary that these trees decompose and, with a certain combination of quantitative and temporal factors and pressure, coal turned out over millions of years - we note, long before the appearance of mankind. In order to learn certain lessons, we sometimes need certain materials, and these materials are provided to us - that's what matters! In this case, people needed a huge amount of easily extractable fuel. It made it possible to make a number of inventions that led man to the so-called industrial age.

Here we come to metals and other types of "raw materials". From my point of view, they are interesting not only for their properties, but for the relationship between their necessity and availability. For example, iron andaluminum is absolutely necessary in mechanical engineering. And yet widely available. But what if, say, gold and silver were abundant on the planet, while iron and aluminum were rare? Then the industry, technology and transport that we have now would simply be impossible.

Another example of Cosmic Planning: almost anywhere on the planet people can find some food and water to drink. If there are no rivers or springs, then it is enough to dig a well right in the ground, and we will have fresh drinking water (which is wonderful in itself). If the ground is frozen, ice or snow is usually available to melt. In addition, entire groups of people are, as it were, specially programmed to live in the most severe conditions. Through this, the physical planet can be fully embraced by the network of intelligence. Since the human kingdom is destined to be the (physical) "global brain" of planetary Life, the next step was required for the implementation of the Divine Plan: the establishment of a peaceful interaction between human communities. This was done through interest in trade.

If the most necessary for human life is distributed relatively evenly across the planet, then this cannot be said about many other useful resources. Minerals, coal, oil, wood. Stocks of all this can rarely be found in one place. Some groups of people have huge deposits of oil, but no iron to build oil-producing equipment. Others have deposits of ores, but no coal to smelt the metals. The rest is clear. Again, this part of the Divine Plan. First of all, such a situation served as a stimulus for the development of our intellect; it was necessary in order to make our life more comfortable. But in the long run,

the most important thing was to make humanity interact and eventually become "unity in diversity."
Let's get back to industrialization.

Seen from a higher level, its most significant achievement is not in the sheer quantity of products produced, but in the fact that, for the planning, production and distribution of goods that engulfed the whole world, it was required that mankind engage and thereby develop its concrete thinking. Until we develop concrete thinking, we remain mostly emotional beings and cannot make very far on our spiritual path. This brings us to another and much more important merit of the age of industry and technology: it has naturally moved into the age of information and communications. But this in itself is not the end goal.

The ultimate goal of humanity in this era is to realize its destiny: to be an integrated "global brain" and the nervous system of our planet. When in planetary events we not only see the "what" happens and the "how", but also understand the "why", it becomes more and more obvious: there is an even grander plan called the "Divine Plan"! But what about those communities that resist interaction and remain isolated? It is very important to note that those who preach any kind of "isolationist" ideology act against the Divine Plan, whether they realize it or not. Evil forces in the world do not want cooperation in humanity. Their strategy is to maintain disunity and division.

We have many examples of stagnant (relative, of course) cultures that have been isolated from others for a long time. But our evolving universe does not tolerate stagnation. When an individual, culture, or even belief system gets stuck and resists growth, and their inner consciousness crystallizes, the energies of change are released! The immediate results of this can sometimes be felt as unpleasant or even severe. But the long-term result is very useful. The same people who had to endure shocks, a much happier life can still await. This reasoning, of course, should in no way justify, let alone encourage, the violence of some people, cultures or belief systems over others. Enlightened people are always trying to promote the progress of their brothers and sisters by personal example and lovingly provided opportunities.

By expanding our consciousness, we are potentially able to create and ascend to happier states of being. We continue to hurt ourselves and others, not because we lack intelligence or guidance, but rather because we still have an underdeveloped energy of Love and we are incapable of empathy (or resist this feeling). Later we will understand what role the other kingdoms in nature play and how they help us fulfill our role in this Conscious Universe. Most importantly, they are necessary steps in the upward spiral of the evolution of consciousness. Perhaps now we can consider in more detail the human stage of evolution, which, of course, is of most interest to us. A spiritual journey (this is also what evolution can be called) is usually compared to climbing a mountain.

Such a comparison is appropriate for many reasons: in evolution it is necessary to make efforts that are rewarded, and mistakes lead to delay; it is easier when you are led and instructed by someone who has already climbed the mountain himself; the more you climb, the more opens to the eye; when you approach the top, it becomes clear that it can be reached by more than one single path (although the closer to the top, the closer all paths converge), etc. Now let me take another analogy. It will not be a spiritual ascent to a mountain, but a journey across an entire continent. Imagine that it begins when we are at a primitive semi-animal stage of development, and ends in our distant glorious future, when we are ready to move into another, higher realm, sometimes called the "Kingdom of Souls."

Let's start the story. The mass of people is on the east coast of a large continent. They are told that they must pass all this vast territory and reach the western shore. Upon reaching the goal, they are promised a great reward. Since they will go on foot, the path promises to be long. It is not a race, but they are expected to keep moving forward. On the way, they will eat fruits and berries, vegetables, nuts and grains and drink water from rivers and springs. With a little effort, they will be able to provide themselves with everything they need. Among them there are individuals who have already had the opportunity to make such a resettlement before. They go up to one settler, then to another, and talk about what a great reward awaits them, and also about the fact that you can save time if in some places "cut way".

But few people listen to them.

So, people gather in groups and slowly set off on the road. Since a huge mass of people was dispersed along the entire coast, most groups operate almost autonomously. Some groups go forward for several days, and then, tired of the road and finding a suitable place, they stop for a while. Others walk past them until they decide to rest. A little time passes, and now the groups have dispersed over a vast territory: some have gone far ahead, while others have barely moved.

Sometimes the groups argue among themselves. Disagreements usually arise between those who follow the call to move on and those who have tasted the charms of a settled life and, having lost interest in the promised reward at the end of the journey, wants to stay in place. Under the influence of opposing energies, a split occurs in some groups: some people continue to move forward, while the rest do not want to leave their homes. It is difficult for those who are ahead, but they are rewarded for their work. They need new knowledge - and they get it. Those who decide to stay in one place spend more and more energy, consolidating and repeating what they already know. Sooner or later disaster inevitably strikes: a flood, or an earthquake, or a terrible hurricane. So in the end, they have to leave too.

Sometimes migrants notice that new people have joined them from somewhere - individuals or groups. This is often resented because the newcomers didn't go all the way from the start, but they're going to get the same reward at the end of the journey. (Does this remind you of anything?) And not only because of this: new people need to be taught what others have learned from their experience. Does this seem unfair? The "old men" prefer not to remember that they themselves were helped a lot: from the gift of life as such to all the other gifts on their way.

In fact, everything is a gift from Above.

Serving a higher purpose and helping others was the least they could do. (But by and large, we humans are ungrateful for the endless gifts bestowed upon us.) Over the very long time that this journey has been going on, almost every group has had a chance to be at the forefront at one time or another. But almost inevitably, people calmed down, became complacent, and the other group got ahead of them. Very often, those who were temporarily ahead convinced themselves (and everyone who was willing to listen) that they were much better than the rest. When at last the first of the groups had ascended the last mountain range, and the travelers saw that marvelous place towards which they were striving, they sent word and, as best they could, hastened the rest so that they too would share with them the great reward. But some are so accustomed to living on the endless plains that they did not believe in a more glorious life and made the fateful decision to stay where they were.

Does this parable seem too simplistic to you? Maybe. But this is how we look at those who are at higher levels and are trying to help us. How many of us resist change (growth)? How often do we cling to the familiar? Consciously or unconsciously, we ourselves choose our path and follow it. And because we are all different—and should be—each path is unique. However, all paths (figuratively speaking) pass through the same rivers, deserts, swamps and mountains. We perceive them as obstacles, but they all serve as necessary lessons for us. When we overcome them, they become milestones on our path to enlightenment.

As intended, our human journey began with the creation personality isolated and self-centered. A personality that we must change and transform - and we will definitely do it. Transformation is achieved through the fire of the mind and leads to the formation of an enlightened Spiritual Being. This process requires a complete reorientation from our focus on the small "I" to self-identification ultimately with the greater life - with the Life that embraces the entire planet! Here one may ask the question: why should we create a strong individuality, if in the end we have to cast it down for the good of the whole? Individuality had to be created in order to develop free will, because they go side by side.

Then we need to learn how to use our free will correctly. Smart at first, then with Wisdom-Love. This process is necessary if we are to become an active ingredient—no less than a co-creator—in the great work of unfolding the Divine Plan. As co-creators, we will use our individual talents and abilities to contribute whatever is necessary for the further enlightenment of humanity. This process requires that we become responsible, learn patience, open our hearts and begin to serve humanity! As individuals, we are just small grains in the universe. But our Soul it is a hologram of the universe and it contains the potential of the All. Therefore, we must release our portion of matter, pushing upward from our personalities and thereby responding to the eternal attraction of our Soul.

We are progressing from the animal group soul to the soul of man as an individual with free will. Then, over time, we acquire the qualities of Love-Wisdom and thus become enlightened co-creators in the Divine Plan of the universe. It has always been a mystery how suddenly (on the scale of natural history), in the absence of a "connecting link", very different and much more developed races of people appeared. Science puts forward postulates that are not consistent with common sense, and our religions generally ignore the problem itself or, in extreme cases, refer to God's providence. By the way, in this case religion is closer to the truth.

It must be emphasized here that even Spiritual Beings act according to the Law. In other words, the means of the physical plane are used to bring about the results of the physical plane. It is interesting that right now, when the prototypes of a new model of humanity are being developed, many people report that they were "abducted" into strange spaceships, controlled by strange (to us) creatures, and that genetic experiments were carried out on them there. There are also documented strange cases of "mutilation"animals, especially cattle, from which organs and sometimes blood have been surgically removed, material that can be used to 'mutate' animals. In addition, new species are constantly appearing in the animal kingdom. (And I would advise watching what happens to the cattle species in the near future.)

It seems that those who accept UFOs as reality tend to adhere to the "alien" paradigm. I would suggest lookingunraveling the mystery "closer to home": in the border area between the physical plane and the next higher vibrational dimension (it is called "etheric plan"). Although these energy dimensions have their own protective "webs" and different vibrational frequencies from ours, they are not impenetrable to those beings who are ordered to help.our evolutionary process. (Later in this book we will talk about these creatures and what can happen with their participation.)

From everything already said in this book, it follows that Life is a continuum, everything is part of something "higher and greater", everything is interconnected and interdependent, everything is unity in space and time. Everything is eternal and moves in a spiral leading to higher levels of consciousness, or enlightenment. What does this mean for us in our human kingdom? How are we connected, for example, with a distant galaxy?

Let's start from the beginning - with the physical body of a person. We know that it is made up of bones, muscles, blood, organs, etc. We also know that these components are made up of cells, which are made up of molecules, which are made up of atoms, which are... well, the picture is clear : everything is interconnected and interdependent. And we're back to the correspondence again: "As above, so below," or, in this case, "As below, so above." We, as individuals, are part of the human kingdom, and the human kingdom is meant to be the global nervous system of the planet, and that is where it is evolving. All the realms (both physical and non-physical) of any planet form the "body" of that planet. This "body" provides the shell for planetary Life. (Just as our body provides a temporary "home" for the Life that lives in us, your true selves and mine.)

In turn, any planet is one of the "energy centers" or "centers of consciousness" in the Life of the great Solar

Being. Any solar system is one of the energy centers of an even greater, more developed spiritual Essence. And this Being, in turn, is also one of the centers of even greater Life, and so on: constellations, galaxies, metagalaxies... All this taken together is our Living Universe! Pantheistic God. And in this regard, I would like to note again: when we look at the heavens, what we see with our eyes is only a vague reflection, a shadow, if you will, of the colossal energies surrounding us and our tiny planet.

The splendor and Glory of the Beings living there correlates with the tiny minds of people, as their gigantic sizes correspond with ours. Proof of? Let's start with the obvious: beauty, harmony, order in heaven. From the course of physics (and our space programs) it is known that in order for an object to remain in orbit, it needs to reach a given orbital distance and speed relative to the object around which it rotates. If it's moving too low or too slow, gravity will pull it in (think of fallen artificial satellites). And if the distance or speed is too great, it will disappear from the gravitational field. (Again, remember the satellites that escaped into space.) Such incidents happen, although the best minds and technologies of mankind are involved in space programs. And are we supposed to believe that countless billions of dead rocks (planets) and suns ended up in their ideal orbits by accident? Not, these harmonious relationships are maintained thanks to the perfect Consciousness of these cosmic beings. But even they have failures, although this happens quite rarely.

We must remember that our planet and solar system, like other solar systems, also grow and develop (in their higher dimensions) with all its unimaginable (for us) high spiritual level. And when they go through their "growing pains", it reflects on us! This may explain many of the eternal myths and legends that we find in all the ancient cultures of the world - myths about giants, gods and goddesses who perform superhuman deeds. These are simplified, personified lower reflections of the vast cyclical energies that have been at work on our planet and in the solar system for billions of years. Although these important cosmic events were dressed in the simple form of fairy tales for not quite mature minds, there was a higher truth in them. Myths and legends are one of the ways to reveal the highest truths to mankind in an allegorical manner.

Another important point: although it seems that "heaven"far away, in fact we are inside them. This illusion of distance is due to the fact that our perception is focused on the physical or other lower planes. On the physical plane, everything looks objective and separate. But on the higher planes, where our Spirit resides, there is no separation (as we imagine it), and all energies interact with each other. For example, astronomers say that our Earth is in our solar system, which is in the Milky Way galaxy, etc. This is the beginning of an important truth. Indeed, in our higherdimensions, we are inside the energy body, the aura of these great Beings (in the ascending hierarchy). Each of us is truly a childstars"!

Or, in other words, we are cells in the body of God. That is why we are deeply affected by these celestial bodies (actually Beings) just as the events that happen to us affect every cell of our body.

It is necessary to understand that the Cosmos entirely consists of powerful energies, or Lives, and we are a small part of the Cosmic Life and are subject to its influence. That is why some of the best minds of mankind throughout history have been studying astrology. (This is not, of course, tabloid astrology.) Using scientific methods and intuition, true astrology is nothing more than an attempt to understand and describe the origin and functioning of the great Life. Although serious astrologers are the first to recognize that their science (or art) has yet to penetrate the surface of Cosmic reality, even now the study of astrology reveals a great deal.

The Life Of The Individual As A Reflection Or Model Of Human Evolution

Continuing the theme of this section (the Universe as our perfect teacher), let's ask ourselves the question: can our life itself be our teacher if we learn to see it from a higher level? What if a person's life from conception to death is actually a model, or map, of human evolution? Orthodox science knows this in principle as the biological law "ontogeny reflects phylogenesis." But, again, science applies this law only to the physical organism. We will apply it also to the spiritual consciousness, which is certainly the essence of the All, and then from this point of view we will try to imagine the future.

We are well aware that the human embryo first repeats the plant phase of evolutionary development, then the animal phase (fish, amphibians, mammals, etc.), and only then takes on a properly human form. This shows us our past evolution and reminds us that our physical bodies are connected to the lower realms. It can be said that during the rest of the pregnancy until birth, the being in the womb is a developing human "personality".

Meanwhile The soul watches and waits for the physical shell to form and for the right moment to be born. The world we live in is not perfect, and events sometimesare not going as planned. Therefore, it may happen that the soul decides not to incarnate this time, and the process of pregnancy ends in a miscarriage.or stillbirth; or the baby may suddenly die. The reasons may be physical (health) or spiritual; the latter are still incomprehensible to us at our level of development. And, although this may be perceived as a tragedy, this being will later incarnate in another body, maybe even in the same mother or in the same family, when conditions become more suitable. In fact, life is never lost!

Eternal Wisdom tells us that the Oversoul (Angels? God? Spirit Guides?) watched over the sub-human, bestial men and women until they were prepared to accept their own Soul each. Then a new stage in the development of mankind began. This momentous event took place millions of years ago. The wave of human life will continue for millions more years, and sometime in the future, most people will leave the earthly plane and move on to what we now perceive as Spiritual Consciousness.

But let us return to that important moment when a new cycle of incarnation begins. A child is born and takes its first breath, the Soul finally connects with a tiny body, and the creature becomes a real Human! To facilitate this event, certain birth rituals are often performed on the child - for example, baptism. Here, by the way, it can be noted that the location of celestial objects at the time of birth can tell a lot to the Wise about where (relatively speaking) this Soul was after it left the previous life cycle, and what it has to learn in the new life cycle that it begins now.

Now let's go ahead and talk about something that is not so widely known. The first seven (approximately) years are spent developing the physical and emotional bodies and brain. At the end of this period, the second seven-year cycle begins - the time of the "Age of Reason" on the scale of the individual approaches. In many religious and cultural traditions, this transition is celebrated (and facilitated) with another ritual. This helps to join the next aspect of the Soul - the true mental body. Now the young Being has a rudimentary ability for abstract thinking and begins an important period of schooling.

Then, after ten years, (as we all well remember) the next component of the whole personality appears - a very important, although still only rudimentary, aspect of love. Its occurrence is associated with puberty, and it manifests itself mainly in physical and emotional love, or in sexuality. And, again, in some societies this significant event is celebrated with a special ritual. (Most of the so-called "poltergeist events" occur when these very strong components of the whole being try to join in.)

Now the Soul is somehow attached to the "sheaths" of our personality: the physical and emotional bodies, the mental body and what corresponds to the "body of love" at this low level. But throughout life, we must strengthen these ties, which we will talk about now. In human communities, it is believed that by the end of the third seven-year cycle, the human Being is already fully formed. With the achievement of adulthood in all cultures, a person already acquires the status of an adult. What people usually don't realize is that the (approximately) seven-year cycles go on and on, the Soul continues to strengthen its position until, after many lives, it eventually becomes completely dominant and "saturates" itself with itself. personality. It is important to understand that the first twenty-one years will form a large cycle, which consists of three smaller seven-year cycles and which will repeat itself on higher turns of the spiral, again following the same pattern (physical, mental, love). Matches within matches!

In other words, from birth until the age of twenty-one, physical expression is primary. Then, for another twenty-one years, our intellect will grow, and the physical will begin to fade. In and after the third cycle, we gain wisdom and a higher form of love. You can observe this in your own life: around the age of forty-two, sixty-three, and eighty-four, important events (changes) will occur or begin. Seven-year cycles are also viewed throughout life - in particular, at the age of 28 or 29, a person usually experiences his "Saturn return" for the first time in his life. (We are talking about the "zodiacal" influence.) It must be emphasized once again that this is typical for everyone, but depending on the level of spiritual development, individuals experience this in different ways.

Because the human realm is clearly still in its teenage years, we are fascinated by the physical world and exhibit other qualities of this age. If we survive and reach maturity, we will revere more higher qualities: intelligence and, most importantly, Love-Wisdom. Our solar system is endowed with this spiritual quality of paramount importance. ("God is love".) It is extremely important to note that at the present period of human historyso many of our supposed "leaders" (in politics, business, entertainment) do not aspire to the highest and most important qualities of humanity. Instead they try to capitalize on everything transient and unreasonable, encourage, protect and thereby glorify power over others, violence and greed. In many ways, this is becoming a "model of behavior" for our youth. They play directly into the hands of the forces of evil! Even in our current (relatively childish) state, we must understand how fleeting glory is. How few celebrities use their fame to help the growth of consciousness, even though we know that the historical figures we revere demonstrated the eternal qualities of wisdom, compassion and love for humanity. Doesn't this mean something? Don't you think that many more people should at least try to acquire and display these qualities?

Returning to the conversation about the life of each of us, let's talk about aging. Why do we (physically) age at all? If all the cells of our body are often replaced by new ones, why do wrinkles appear and the body gradually loses its former health? Besides, if our intelligence were completely dependent on the brain, then wouldn't we start losing our mental abilities as soon as we grew up? In fact, our knowledge and, more importantly, our wisdom increase with age. Could it be that the gradual loss of sexuality from a relatively early age contributes to the development of our consciousness? Maybe it is then that we concentrate all our attention on what we incarnated for? That is, on expanding and raising our consciousness, increasing intellect, wisdom, the power of love. Precisely because Perhaps, losing the physical, we begin to listen to the instructions of our Soul and give more and more energy to spiritual aspirations? After all, it seems that we actually become wiser and more sensitive as we age.

Older people usually have a more developed taste for

music, art, for what we call culture, for more refined and higher qualities of life - qualities that resonate more with the spiritual realms (correlation again). Most of us do not begin a contemplative life until we have outgrown the entertainment and other energies of youth, unless we are talking about a very "old soul" who demonstrates wisdom and compassion even in (physically) young. Doesn't all this point to the fate of mankind in the future? No, it's not at all about the fact that the body will be ugly and wrinkled. I mean the maturity of values: there will be a gradual increase in the proportion of people who are more polarized in the mental and higher bodies (which we call spiritual) and less in the emotional body (the body of desires).

As for our physical bodies, they will become even more beautiful and perfect. But beauty will no longer be identified only with the sexual attractiveness of a person, as it is now. Our physical beauty will last until the individual age corresponding to the evolutionary age of the human kingdom. In other words, when the human kingdom is about halfway to its destined spiritual growth, people will reach the peak of beauty not in youth, as it is now, but in middle age. Inner beauty, which increases with age, will be manifested in the beauty of appearance. It is said that even now some spiritual, or angelic, Beings continue to look young, having already lived a significant part of the life given to them.

This is also observed in the plant kingdom, which has undergone a great evolution (insofar asThus, we have shown how the typical individual life of a person repeats and demonstrates the past evolution of our spiritual consciousness and how it indicates the path that lies ahead of us. Now we can look at the entire family of humanity and trace human evolution from the animal stage to the present. Stages of the evolutionary path of human consciousness:

a) Hunting and gathering

b) Military affairs

c) Agriculture Crafts

d) Tradel Industry

e) Information and Communications

The science of anthropology argues that people began their journey in many ways like animals: there were families, extended families, and groups of families (clans or tribes). They worked together, getting food for themselves, looking for suitable "camps", supporting each other, etc. As more and more people searched for food and suitable places to live, competition arose, followed by aggression; it became clear that the strong had more chances to survive. This is how the warrior class was born.

In the end, some people learned to grow their own food andrealized how much more convenient it is than looking for her. At some stage, they began to capture and tame animals in order to have meat, milk, skins, etc. This allowed families and tribes to settle in one area and freed them from the need to constantly move to get food. The need (which eventually led to the ability) to make various things was a logical consequence of the beginning of the formation of society and the development of agriculture. This is how crafts and arts appeared.

Naturally, the neighboring tribes and clans began to trade and exchange goods with each other, and then the class of merchants gradually developed. A universal medium of exchange, or money, was required. As human intelligence expanded, better and more efficient ways of producing goods arose; this process culminated in the so-called industrial age. More and more knowledge was required, as well as the means toacquisition, storage and exchange: this is how the current information age began. And so we come to the first major rung or stage of the Divine Plan for the human kingdom! Now we are starting to build a "global brain"! It is necessary to realize the great significance of this most important step. Soon the planet will be able to function as a whole Being! This is what frightens the forces of evil most of all, and therefore they stubbornlytry to support separatist thinking among the peoples of the Earth.

Before moving on, let's look at the good andbad sides of the stages described above people at these stages of evolution. The Hunter-Gatherer stage gives birth to individuals (and social institutions) who are looking for new sources of material resources. They can become pioneers and trailblazers. Those who have not reached development in this category become thieves, swindlers, swindlers, etc. The Warrior class develops into a police force and an army, which must protect society, acting according to its laws and under its supervision. However, human history is replete with examples of cruel lawlessconquest wars. There is no need to mention all this here.

At the agricultural stage, developed people respectfullyrefer to the earth and to all life that is an integral part of the ecosystem. Therefore, they cultivate the land, extract minerals, use water and other resources wisely and understand that if everyone acts with intelligence and good intentions, if everyone shares with each other, there will be enough livelihood for everyone. If the economy is conducted ignorantly, stupid and greedy, we get just everything that we have today: "factory farms", monocultures that deplete the soil, environmental pollution - and many, many other problems.

It seems that craftsmanship and genuine art are now becoming rare. But new energies come to the planet, and when humanity begins to act on a higher turn of the evolutionary spiral, these skills will not only be revived, but will also increase and will be appreciated.
Much of what is now passed off as art is not. After all, true art is always a reflection of cosmic harmonies and proportions at a lower level. **Trade** conducted ethically is the recognition of our interdependence; it aims to create commercial relationships where everyone wins. It contributes to the development free enterprise that encourages people to make the most of and develop their talents and abilities. Money should be used as a medium of exchange, allowing a person to acquire everything necessary for life and start his own business. When capital is used primarily for manipulationothers and personal enrichment, and there is no benefit to the

common good, it's just a crime! Remember, unfettered capitalism should eventually theoretically lead one person to have everything and the other person to have nothing. Free enterprise and capitalism are not the same thing! Greed is a disease and too many people are infected with it. We will talk more about the perniciousness of materialism in the next section.

The positive side of industrialization is that it allows the production of sufficient quantities of everything necessary for the life of mankind. In addition, over time, thanks to industry, people even have some abundance, allowing them to have free time and spend it on expanding their knowledge. In this way, people become more and more intellectually developed, and this is, of course, an important factor in building an integrated human kingdom. We are all familiar (including from our own experience) with the inhuman consequences of excessive industrialization, including environmental ones; it is not necessary to list them specifically here.

Information and communications in elemental form have always been available even in the lower kingdoms, and the history of knowledge and communication is considered to form a significant part of the history of evolution itself. But only now information technologies are beginning to take their rightful place as the main activity of mankind. And, although much of the incentive to expand knowledge and communication was (and still is) based on personal selfish motives - such as greed, the desire for dominance, pride, etc. - ultimately all this for the benefit of. In time, the planetary communication system that is now being developed will be used more and more for the benefit of all the kingdoms in nature that make up Planetary Life. Eventually there will be unrestricted global interaction, i.e. each person will be able to communicate freely with any other person on the planet. Although this is a matter for the future, even now one can see its benefits for humanity. With the help of the Internet, people with similar interests are in contact with each other, regardless of political boundaries. The "Age of Aquarius" is characterized by the emergence around the world of informal groups created as a result of such communication.

This is a necessary component of the Divine Plan! Therefore, the dark forces have always tried and will always try to control, restrain and in one way or another interfere with the ability of people to interact freely. This must not be allowed! Cultural exchange, tourism and trade on a fair basis - all this also greatly contributes to the rapprochement of people and the growth of mutual understanding between them. If we aspire to become citizens of the planet and to interact in peace and for mutual benefit, we must understand that this is possible only if we acquire the quality of responsibility. (As we receive more Light, we develop the "ability to respond" properly. This is true spiritual responsibility.)

It is often said that people "do not take responsibility" for the consequences of their actions. Responsibility is not something that can be take or not take. By definition, we are always responsible for our thoughts and their consequences. Let's once again - from a different angle - look at the development of an individual human individual, comparing it with the evolution of mankind up to the present time.

When the Cosmic Light descended deeper and deeper into matter, ordarkness, the "Rays" of this pure Spirit, or Divine Monad (someone would call it a "spark of God") dissipated, penetrating into the densest matter - into what we call the "kingdom of minerals." Then the work of liberation began, that is, the planting of consciousness into a part of the unconscious life. After billions of years, Light created a "pre-consciousness" that grew as it moved upwards, encompassing the plant and animal kingdoms. Eventually, when the Light received the guidance of the Solar Angel or Soul, it became a member of the human kingdom.

Here is what is important to remember: in essence, we are the immortal spark of God, or the Cosmos! But once upon a time we were only formally human beings, living mainly by animal instincts, and our Soul had to make efforts to guide us and develop our true humanity over a long period of time. Therefore, when any of these beings (that is, us) begins their incarnations on the physical plane in order to go through the school of life, this person begins his journey from a relatively primitive infantile stage. He is still much like an animal and acts like a hunter-gatherer, following the path of least resistance, that is, living only on what he can get for himself. This continues as long as he is in the society of hunter-gatherers. But when he begins to incarnate in a more advanced agricultural or trading society, where goods and services are acquired through barter or in exchange for money, such behavior becomes unacceptable.

At this stage (at the beginning of evolution), people have not yet developed a conscience, and as they grow older, they often come to the idea "who is stronger is right." Even today, "young souls" (those who have had few physical incarnations) are often in this "childish" state. They live only to satisfy their desires. We also know that some individuals, even those with a developed intellect, still remain essentiallypredators and get what they want by the most primitive means. Society should take this into account when organizing the work of the judicial and correctional systems (and other institutions). We need to try to find ways to plant a new consciousness in a person, and not just put such people behind bars along with others who are at the same early stage of evolution. Everyone is well aware that this is of little use.

Please don't misunderstand me: there is nothing wrong with the primitive hunter-gatherer lifestyle. It's just that we all need to take advantage of the opportunities given to us to move to higher levels of the school of Life on the planet in order to fulfill our Divine destiny. Why? Because the evolution of man to enlightenment, as well as the responsibility associated with it, are planned by spiritual Mentors, or Hierarchy (or God, if you like). If we get stuck at any stage of our spiritual evolution, we will obviously never fulfill our Divine destiny. The next step is the beginning of cooperation, but so far only for the sake of benefit for oneself.

Since life is often threatening and chaotic at this level, we begin to adhere to certain laws and maintain order. Butat this stage, people are usually more concerned with getting others, rather than themselves, to be law-abiding and disciplined. Power, strength and control are still highly valued. After many incarnations, having accumulated a lot of experience, having made a lot of effort (and having gone through a lot of pain), a person gradually learns that it is much more pleasant to be among people who demonstrate such qualities as responsibility and good will, and that in this for us, perhaps, there is some message. It is at this stage that we begin to open up for contact with our Soul, and since our Soul is a part of the One Soul, we acquire a new quality - "sympathy" and As a result, we begin to show some concern for the well-being of others.

We no longer live by our own interests. Altruism begins to flourish! After many incarnations, the good will gradually becomes the will to good. This means that now it is actively operating at the level of intention and becomes our "second nature". As already mentioned, this is a very important moment in our spiritual evolution! There is nothing surprising in the fact that religions that appear at different periods of history usually correspond to the level of developing consciousness. Primitive religions usually concern themselves with quite physical things - for example, animals and parts of their bodies - and sometimes even they try to call upon the elementals, or nature spirits of the lower astral (emotional) plane. Each tribe has its own gods. They are connected with the earthly and "mundane" themselves, can be cruel and sometimes even require living victims. On a higher level, early religions can help physical and psychological healing and open people's eyes to the fact that there is life and Spirit or Soul in everything.

Then we have gods created in our own childlike image. First of all, these are jealous deities who want to be served and worshiped. They control us through fear and guilt with the help of unshakable, simple prescriptions thatimposed by intimidation: infidels ("them") are promised terrible punishments in the afterlife; but the elect ("us") awaits a blissful eternity. Emotional Rules! At this level, religions are sometimes usurped by those in power and "God" only complements the rulers: He favors a certain gender, race, nationality, and someone's current political and economic ambitions (doctrines). It happens that a person, having become a ruler, appropriates the status of a god or divine qualities.

We are well aware of the terrible crimes committed in the name of fear-based religions. On the other hand, the fear of such religions led many people who were characterized by antisocial, criminal behavior to the first stage of ethical behavior. But we continue to evolve, our minds become more active, and some beliefs, accordingly, more and more meaningless. If there is a God, then God must be better than us, not as bad or worse. Emotionally based dogma is being questioned more and more. There is less and less faith in eternal heaven or hell, because it becomes obvious that a truly loving person cannot enjoy life while others suffer endless torment, no matter how much they have sinned. And it's not just that: the purpose of "punishments" and the transferred pain is to end something, to teach us something so we can grow longer. But endless suffering cannot serve this purpose or any other.

Understanding this, a person gradually moves away from a religion based on feelings of guilt and fear, towards religions based on Love (and which are intellectually more healthy). The focus is shifting: if earlier all efforts were aimed at appeasing God and thereby saving one's own skin, now a person begins to worry about all creatures. Conscience begins to develop. And all this time we are more and more adapting to civilization. After many lifetimes, we begin to develop true culture. Although we may not realize it, we are now becoming, in a sense, spiritual beings.

And so we come to the next stage, when we often question religion, and sometimes even reject it for a while. We can spend more than one lifetime developing the lower mind but moving away from the control of the emotions. Often at this stage, religion becomes, as it were, a science, or, better, "scientism." The concrete mind (or, as they say now, "left brain" thinking) develops too much and takes over the personality. This mind is convinced that all answers can be found in the material realm, simply by taking things apart and studying their constituent parts. At this stage, the lower mind becomes the "killer of the real" (as it is called in the Teachings of Wisdom), because it is unable to see the higher, abstract reality - true spirituality, - and denies its existence. Therefore, those those who are focused on a particular mind often find unfounded the truths of those people who are capable of operating at higher levels. Intellectual conceit is a trap that many have fallen into at this stage.

Or, on the contrary, we adhere to the "right hemisphere" and become more mystics. As we grow wiser, our gods become more like our parents: we expect them to answer reasonable calls, and we trust that they care about our well-being and the well-being of others. We understand that all people have to learn lessons ("What will be, will not be avoided") and, in the end, we receive them by fully experiencing the same pain that we have caused others.

Then, after many lifetimes, a larger picture gradually opens up to us. We are beginning to understand how impudent it is on the part of a weak little man to think that he has at least begun to comprehend the Creator of the Universe! In terms of the level of consciousness, we are much closer to insects than even to the lowest of truly spiritual Beings! Finally, we gain humility and a sense of proportion. And only then can one begin the long ascent to Divine Wisdom. It is at that time that we comprehend very important things: everything is part of an even greater whole; there is one unchanging all-encompassing Principle; the universe is an evolving hierarchy, and "Great Universal Design" (as some call it). And we are an important part of it!

People who have reached this stage of spiritual growth—that is, responsible, compassionate, altruistic, exercising an intelligent, efficient will-to-good—are regarded from above as the "New Group of World Servers." They are working for a higher purpose, an evolutionary purpose, whether they know it or not. (Many don't know. But people with these qualities really serve the Divine Plan.) A little later we will talk about the further stages of the Path of Discipleship. From time to time beings are born among us, bringing newmessage showing us the next steps of our spiritual growth. We kill them, and when a lot of time passes, we only gradually and reluctantly accept some of their teachings.

But the dark forces usually manage to build some kind of religious institution around new truths and to a large extent emasculate the Spirit from them, diluting them, dogmatizing and politicizing them. There is a kind of gravity in the human realm, an urge to descend to the lowest common level, and if this is not resisted, the result is always disastrous. We see this process repeated over and over throughout the human wave of life. Just listen to those in positions of power (whether secular or religious) and you will note with sadness how rarely do they demonstrate even a fraction of true wisdom, let alone more.

But this situation is about to change with the advent of new enlightened Souls. The inspired individuals who started the great religions did so to shed light on the path that is open to all of us, and all true religions will continue to guide us. A big problem arises when a church becomes compromising and conceited and begins to believe that it in itself is the ultimate goal. When a church leader says, "You just have to come to my church and you are saved. I have known the truth, the whole truth and the only truth!" - this person hinders our spiritual growth rather than helps! It is simply an indulgence of that weakness that we all have: the desire to "be holier than others." Such a perverted way of thinking has already led and now leads to bloody religious wars and persecution of non-believers.

Let me explain my thought: religions have always been, are and will be a strong and necessary means of enlightening mankind. But, as in everything else, we must be picky about what we accept as universal truth. Spirituality comes from what we really are: the Spirit. Religion on the other hand, they are collectively shared beliefs about reality. Our Soul, the higher Self, the "Kingdom of God within us" is our only reliable guide, and we must willingly follow his guidance.

Before we finish this section on the Universe as a Teacher, it is necessary to pay special attention to one point: all problems in all kingdoms of life, in all spheres of life, are surmountable and ultimately solved only by raising consciousness! Due to Spiritual Enlightenment and Love. This is one of the deepest truths that a person can know, and the truth that he must definitely think about and understand. All other attempts to solve the problems of mankind are only temporary measures.

No "sticks" and "carrots", whether it be material well-being, good health, all the benefits of a happy life - or punishment,coercion, guilt, fear, etc., by themselves have never stopped and never will stop "inhumanity between people" (Robert Burns, "Man is born to mourn"). But they lead to a gradual riseof our consciousness, as a result of which a person does more "right" and less - bad. And, again, only the growth of consciousness, both at the individual level and at the level of the entire human kingdom, will lead to a just and peaceful life.

Beings acting from the level of the Soul do not harm others either by their actions or by their thoughts. Take any scenario of human suffering, and when you analyze it, you will see that it was caused by ignorance or stupidity, directly or karmically caused by the action of some aspect of planetary lives. Even so-called natural disasters teach us something. In other words, the life cycle of the universe is the time it takes to raise the consciousness of all Life in the universe to perfection. Or - to the Universal Enlightenment.

This does not mean that we have to wait billions of years for our suffering to be relieved. With the growth of individual consciousness and understanding that leads to right actions and right thoughts, we will increasingly enter into a state of joy. True spiritual teachers always rejoiced, even when living in the most difficult conditions! Let us repeat once again: always from matter (external) - up through the mind, or consciousness (quality) - to Love-Wisdom (Spirit, or Life). This is the Path of Enlightenment.

We see it both in our lives and in the evolution of our planet. If it were given to us to see the complete picture of the universe, then we would see it in the evolutionary return of the entire Cosmos to its perfect Source. And He follows the same path. This is the true "liberation of matter"! It is liberated, or rather, re-spiritualized through the eternal life cycle of the universe. This is the ultimate meaning of life. This is the Divine Plan, and we are a part of this process, and a very important one at that! Someone will ask: "Why don't the teachers of the human race simply tell and show us these higher truths, so that we never doubt them - so to speak, they won't be inscribed in heaven?"

There are several reasons for this. The main thing is that we would not have learned to know then, would have become even lazier than now, would continue to follow the path of least resistance and therefore would remain dependent children (in the spiritual sense) even longer. Yes, high truths are often distorted to one degree or another. Therefore, we need to constantly expand our minds, which is the path to wisdom. There are many phenomena that can be called mysterious. They can be interpreted (or ignored) in different ways: it depends on the degree of enlightenment of a person.

Therefore, people who do not want to change their beliefs assure themselves that events that go against their views do not actually occur. Some call it the "law of disorder," others call it the "uncertainty principle." The teachers of humanity have always said that as we progress we will see that there are many levels of apparent reality. We must strive for a higher level, not only to expand ourselves, but also because our higher self constantly evaluates, Eventually we reach the stage of wisdom and indeed we begin to see the perfection of the Divine Plan and the great Truth, opening in the incredible beauty of our mundane experience. And then we begin to understand: it was "written in heaven"!

Throughout history, mystics in all parts of the world, professing all kinds of religions (or none at all), have experienced this insight, and they are constantly trying to explain it to everyone else. Good. If we are a part of this Universe, this huge Being, and are immersed in an ideal environment for learning (cognition), why don't we grow, evolve much faster? Why do we "miss"for that? It seems that many of us are quite satisfied with ourselves and would like to remain the way we are.
Now we will talk about this.

Where Have We Been (And Why Are We Still There)

I feel sleepy and I lie down to rest. I think I dozed off, but suddenly woke up. For me, this day is very important. Our tribe roamed the area looking for a place to find food. Yesterday one of our trackers came back here (where our tribe is temporarily located) and said that he saw a family of animals large enough to provide food for the entire tribe, but not so large that they are very dangerous and difficult to obtain. Today he will lead the warriors there, hoping that the animals are still there.

Why is this day so important to me? After all, this happens quite often in the everyday life of any tribe. The search for food is what the whole life of our tribes revolves around. This day was special for me, because for the first time I was allowed to participate in the hunt - I finally became a warrior! Every youth of every tribe can't wait until he becomes big enough and strong and agile enough to be takenfor such a hunt. For as long as I can remember, it seems that I only dreamed about this, preparing for this day. What does "hunt like this" mean? And why do you need to be appointed a warrior? I'll tell you why. The entire tribe is constantly engaged in hunting or gathering. Looking for and collecting food while wandering nearby is a common thing. But to hunt animals, moving away from the camp,This is completely different. It's all about danger: in the wild forest we canunexpectedly stumble upon unknown animals. Or even worse, the warriors of other tribes who can also hunt in the same place.

The results of such meetings are unpredictable. Sometimes, barely noticing each other, groups of hunters simply disperse in different directions without making contact. Sometimes they can come up to each other and exchangegreetings. But if one of the tribes experiences severe hunger, which often happens, then the meeting becomes a matter of life or death. When one tribe has found a good place to hunt, or when they have already killed animals and are on their way with prey, warriors from another tribe who meet them can attack them, maim someone or even kill them. So it happened at the last hunt (then two of our soldiers were crippled), that's why, so to speak, I was "pushed forward". If I show myself well, I will be accepted into the warriors for real.

But if this is my first outing as a warrior hunter, then how do I know all these things and why do I feel so confident? It's just that I've been preparing for a long time. From early childhood, I heard many times howthe men talked about hunting. And not only the warriors themselves, but also former warriors, and those who were soon to become warriors, and those who only dreamed about it. It seems that they did not talk about anything else at all: they boasted of past successes, lamented past failures and argued about how they should have acted in order to turn out differently. Endless strategies and tactics for any situation: how to sneak up on an animal, how to kill it and bring it home so that warriors from another tribe do not take it away. This was discussed in great detail, because all this must be known in order to survive. No wonder I feel quite prepared. Everyone should be prepared for the hunt, because lately food has been scarce and the tribe has been starving. We needed to get food.

And then came the day of the hunt. We warriors come together (I love the "we warriors" thing so much). We're on our way and the hunt begins. Silently following the tracker, I think about how the hunt brings the whole tribe together and how everyone plays their part in it. Other strong men remain in the camp, ready to repel any danger from outside while we are away, or to help if we are pursued (these are our reinforcements). Women, old people and children help us get ready for the journey, encourage us, and when we return, they will greet us with wild delights and arrange a real feast. Well, and, of course, girls. I have often noticed that the most successful warriors are liked by the most beautiful girls. So today I saw that the girl I like and whom I would like to like behaved differently with me. She somehow especially tried when she wished me a successful hunt and expressed hope, that I will return safe and sound. But her smile and look said even more...

And now we have come to the right place. We stretched out in a line, as previously agreed, to locate and surround the prey before we ourselves caught her eye. And then it started! We saw some wild pigs just as they spotted us. While I hesitated, not knowing what to do, more experienced hunters surrounded one pig and all together tried to knock it to the ground. But it was not easy, because the pig wanted to live just as much as we wanted to eat. I "danced" around the fight, trying to cover any gap through which the animal might escape. Exactly thisI was supposed to do according to our plan. Finally, after many futile attempts to escape, the pig became exhausted, one of our strong men tightly grabbed it, put it on himself, and with this squealing burden rushed to our camp.

And then something happened that we least wanted. We spotted another squad of warriors. Obviously, they heard a noise and ran towards us. Their forces were fresher, and it cost them nothing to defeat us. In addition, our tracker noted that some of them were from a tribe called "bears" by our old people for their strength and cruelty. We prepared to fight and at all costs to save the booty so hard-won for us. Excitement, fear, anticipation, anger - all mixed up. I remember quite vaguely what happened next. The two squads clashed, waving their arms and legs, kicking, fighting with sticks and fists. I took a lot of hits and hit relentlessly myself. Our pig revived and slipped out of the arms of the hunter holding her in a commotion. One of "bears" grabbed her and tried to escape.

Although we were tired, we were not going to give up. We chased after him, overtook him and threw him to the ground. The pig broke free again, but this time it was grabbed by one of our strongest and fastest men. Encouraged by this turn of events, we surrounded him, trying not to let a single "bear" come close. The fight continued, but we did not give up. Finally, we were not far from our camp and, hearing a noise, they ran to help us. We have achieved our goal!

I have never experienced such an uplift in my life. Everyone was shouting and waving their hands. And then it's better: "my" girl ran right up to me and jumped with delight. I remember that then we hugged. I was sweaty, dirty, out of breath, and she hugged me! I was delighted!

And I woke up.

Woke up! So it was just a dream? Can't be! Everything was like in life: just as bright, lively, emotionally! I don't want to forget it. Such a vivid and lifelike dream must mean something important. Interesting... well, I might think about it some other time - football is starting and I won't miss this match for anything in the world! But what about some kind of football match when we talk about the most important thing in life, about universal truths? Answer: The fact of the huge popularity of the so-called sports games tells us a lot about where humanity is now on its evolutionary path, and also indicates to the wise what we need to overcome.
There is, of course, nothing wrong with sport in and of itself.

Generally speaking, playing sports is a good way to release physical and emotional energy and, of course, it is much better than war (which has actually always been a sport for aggressive people). In our time, when wars have become too terrible to be commended, it is no coincidence that the sport began to gain more and more popularity. Although competitive sports are generally quite harmless, this is an example that shows us not only the force of the "attraction" of matter that we have to overcome, but also how susceptible we are to the influences of ancient thought forms in the aura of the Earth, or, in other words, to memory of ancestors. (And there are many other examples that are not so harmless.) We must also understand that people lived in tribes and hunted for millions of years, that is, much longer than the period of agriculture and trade lasted. Moreover, the very survival of a person depended on the success of the hunt. This explains why such thought forms are much stronger than those that appeared much later. As discussed in the previous section of this book, there are people who, even now, are just beginning to grow out of these initial phases of the evolutionary process. Sport is just one example of how strong and emotionallyour past holds on to us.

If you don't believe that sport comes from ancient thought forms, let's do an analysis. Any sports game usually begins with the fact that groups (or, in the simplest case, pairs) of competing people gather. Often, clubs, rackets or bats resembling clubs or axes are used in games - as well as balls or similar objects (the size of a small animal or bird). These objects need to be passed over or around some obstacle, get into the "basket" or "gate", hammer them with a stick or cue into a hole, etc. Does this not resemble the process of catching and hammering hunting prey and delivering it "home"? In this case, you need to outwit or overpower another tribe ... that is, another team. In big sports, the opposing team is always from another place, only children play sports games "among themselves".

It is curious that the Americans to this day call the soccer ball "pigskin" (pigskin). Is it necessary to have great imagination to see in this ball the pig from my dream, for which two groups of primitive people fought so fiercely? (Especially when it comes to American football.) As I have already said, most of the "sports games" are generally harmless examples of the influence of ancient and not very ancient thought forms, preserved in the earth's aura, associated with obtaining food.

But there are many such "remnants of the past" that can be very dangerous. Suffice it to recall the bloody wars for land that are still taking place to this day. Peoples are fighting for the right to own the territory where their ancestors lived thousands of years ago. I know this is a touchy subject, because there have been occupations and forced displacements, and some peoples do have a legal right to demand the return of their native land to them (of course, everyone has the right to a decent living space). But this attachment to the "soil", when it is taken to an extreme, prevents a person from looking "up" and concentrating his efforts on the path of ascent to our true Motherland.

Throughout our lives, the Soul may now and again want a person or people to move - so that they communicate with other people and receive new lessons. Remaining in the same place for a long time, the people come to stagnation, because here all the lessons have already been passed. No wonder humanity is becoming more and more mobile and *global* community. Enlightened people take advantage of new possibilities of freedom to diversify their experience and learn something. Returning to the question of how sport fits into the larger picture, there is another important point to make. For an object to fly (this is known to any pilot), the lifting force must overcome the force of gravity.

The same is true when reaching spiritual heights. As with an airplane, there are forces that want to lift us up and forces that want to keep us down. The energies that lift us to spiritual heights and move us forward into a new consciousness are divine planetary guides, as well as our own Soul. They are opposed by forces that want to keep us down; some of them are obvious and are called "the forces of evil", others are not so obvious and therefore more difficult to overcome. The energy of matter itself has very low vibrations (speaking in a spiritual sense), and in order for the higher kingdoms, including man, to progress, this property of matter must be overcome. Much of what happens in the physical world is a "struggle" between Spirit and matter, which manifests itself in man as a struggle between Soul and personality.

As discussed in the previous section, the universe is our teacher. Therefore, be especially careful about symbolism: it cantell a lot. The heaviest level of matter is the mineral realm, which is essentially unconscious and motionless. The next, less heavy kingdom, and with the beginnings of consciousness, is the kingdom of plants, which have limited mobility. Next comes an even lighter realm with even greater consciousness and mobility - the animal kingdom (the class of birds is also associated with the realm of the devas). And, of course, the human realm (as a whole) is the lightest and most mobile of all the realms on the physical plane. Many do not realize that the higher or spiritual realms are so light (and enlightened) that we cannot even physically feel them, and of course they have already achieved what we would call almost unlimited freedom.

We also know that the plant kingdom gradually destroys and consumes the mineral kingdom, which, in turn,absorbed by the animal kingdom (and the animal form of our human bodies). These physical processes correspond to the rise of consciousness in the higher realms. For example, when we (or members of the animal kingdom) eat plants, our higher energy is actually beneficial to the plant kingdom. Another thing is when people eat animals, because the energy of the latter is often strong and coarse and, acting on a more sensitive human constitution, has a coarsening effect.

Always look in terms of energies! Therefore, most often the use of meat is not encouraged in spiritual practice, and if it is allowed to eat meat, then the meat of the lower, less cruel classes of animals is recommended - fish, seafood, but not the meat of carnivorous mammals. And therefore, by the way, we humans thermally process meat for food, using the power inherent in fire to expel some of the gross animal energies.

Let's talk about goals higher stages of kingdoms. The main goal of the mineral kingdom is to acquire the quality of organization. Look at some beautiful crystal and think about how high the level of organization must be in order to achieve such perfection. Interestingly, the highest stage in the evolution of the mineral kingdom is considered to be radioactivity, when the form is no longer able to support the life that dwells in it - and again we are talking about a high degree of freedom. Something analogous to such transformations on the physical plane also takes place in the subtle realms. When the consciousness of the most advanced minerals gradually rises to the level of the "first floor" of the plant kingdom, the essence of their soul is transferred to this kingdom.

Then begins the journey to a new level of consciousness. As the simplest plant life develops into higher and higher forms (including trees, often called the "lungs of the planet"), the (group) soul awakens. In the end, there comes a climax when the "soul" can manifest itself through the beauty of flowers: freedom is expressed through their ability to radiate smell and color, which attracts more developed insects, as well as birds and people. We the people honor flowers when we use them in our most important rituals and acknowledge their subtle healing power when we give them to the infirm.

The goal of the plant kingdom is to learn to feel. Gradually, this will lead to elementary emotions and desires, when the energy of the soul passes into the animal kingdom. The wave of life is moving up through the animal kingdom, complexity and the mobility of organisms is increasing; at last the wave reaches the highest standard of living in this kingdom—domestic animals. They have the greatest freedom of movement, while they want and can accompany a person everywhere. Therefore, when taming an animal that can be made domestic, we change the animal spirit in it to "pre-human", and to some extent it begins to consider itself one of us.

The goal of the animal kingdom is to gradually acquire emotions and desires and then develop these feelings to an almost mental level. (We know that some pets are quite intelligent.) Because this realm begins with single-celled beings, these processes take vast periods of time. Well, that's all great, but what's in it for us? The problem for humanity is that while all kingdoms are striving for enlightenment in the long run, the strong and gross energies of matter, the inertia of matter, are dragging us down. In a word, the problem is materialism.

Mankind does not realize how strong the influence of theseforces on our kingdom and how susceptible we are to them. Things (Matter) have blinded most of us.

We are so deeply immersed in their spell that we no longer notice them. It is for us as water is for fish. It is said that "the love of money is the root of all evil." And it is true. The love of money (material) is indeed the root of almost everything bad in the human world. The three "M" - materialism, monetarism and militarism - are not evil in themselves and even play a necessary role in human evolution. The only problem is our over-attachment to their energy. And the bad thing is that our public institutions support this mentality.

Here it must be emphasized that gross matter gives us another, even more dangerous illusion: at the level of matter, everything seems to exist. separately. Most often, when we are in the human kingdom, we do not realize that we are a part of it and are connected with everyone else in it, as well as with everything that is in all other kingdoms, on the entire planet and even in the entire universe. Once we understand this, there will be an end to wars, crime, and deliberate harm to others. We will begin to adhere to the Golden Rule: to treat others the way we want others to treat us. (We'll talk more about this soon.)

We must understand that the human kingdommust also work to gain freedom, but we are not free if we cling to the material! Throughout the history of human evolution, all spiritual teachers have emphasized the need to overcome our attachment to the material. Indeed, we cannot "serve two masters." When we focus our energies on material things, we deprive ourselves of the ability to support the growth of our consciousness.

A person gains maximum freedom when we take control of our life and free ourselves from the spell of matter, when we begin to act at the level of our higher bodies under the direct guidance of the Soul. In doing so, we eventually consciously enter the path of spiritual discipleship. Then only we, in fact, become people in the full sense of the word!

"Material" is not only "things" on the physical plane that can be heard, seen, touched, tasted, smelled. There are higher correspondences of matter at the lower levels of all planes. Take, for example, the astral plane: there our desires arise, associated with material wealth, money and physical (including sexual) sensations. At the lowest level of the mental plane, we figure out how to satisfy our greed and sense of superiority, and convince ourselves that there is only the reality that we physically experience. It is time to stop wasting so much energy on these low, relatively material levels!

It is well known that very often people who have been saving all their lives riches, become very unhappy and devastated with age and end their lives as simply miserable creatures. It happens that the life of their children also fails, because along with the money they inherit distorted values. One can judge the evolutionary status of a wealthy (or powerful) person by whether he is only trying to maintain his privileges and capitals, or whether he is inclined to care for the less fortunate and advocate a more just order that provides everyone with equal opportunities to use earthly resources.good things. Really happy are those rich people who see the

Light and free themselves from the shackles of materialism; such often become great philanthropists.

Highly developed beings well said: "To whom much is given, from thatmuch will be asked." We need to constantly evaluate what we are spending our energy on. Our way of life not only influences our immediate environment, changing it for better or worse, but also shows the mentors of humanity whether we are learning some lessons for ourselves and whether we are ready to take on even more responsibility.

Therefore, many spiritual seekers prefer to live modestly and unpretentiously and consider any environment ranging from asceticism to modest prosperity to be worthy. After all, true beauty is simple and unobtrusive. This in no way means any special nobility of poverty. We must strive to be the Master of our lives and not be a slave to either money or poverty! The key here, again, is the ability to distinguish and a sense of proportion when setting priorities.

Free Will Individualization

We have already said that the distinctive quality of the human kingdom is free will. In the animal kingdom, there is one group soul for each animal species, and therefore the behavior of representatives of one species is quite similar and typical. We humans are completelyunpredictable, at least until our personality becomes whole and then aligns and merges with the Soul. We should sort of invite the Soul into ourselves and learn to follow its guidance. Until that time comes, we will reap the rewards of our inability to use free will, experience pain and suffering, continuing to make destructive choices over and over again, until we finally realize that no one should lose in life. And it will be much better if you act wisely and make efforts as a group, that is, manifest the qualities of the Soul.

Free will is required in the early stages of human experience in order to build a strong individualized personality. Then in order to integrate the components of personality (physical, emotional, mental). And then - in order to align the whole personality with the Soul. To become a whole, aligned personality, demonstrating the qualities of the Soul — this is the goal of a person at the present stage of evolution! All this is necessary in order to acquire the unique qualities that will later enable us to fulfill our special role in the Divine Plan. If a person is in contact with his Soul, then already now we perceive him as an integral personality.

In the previous section, we talked about the typical human life cycle as a reflection of the larger life cycle of the human kingdom on the path of evolution. From a global point of view, it is interesting to observe how states and other public institutions often follow the same life cycle model as a person. For example, young states (or states led by spiritually undeveloped leaders) usually behave like young people: they are passionate about physical (military) strength, "prettyness" (appearance) and the accumulation of toys (gross national product). In contrast, developed countries usually value wisdom, art, and true beauty more. In other words, for them, the qualitative side of life is in the first place, and not the quantitative one.

It seems that now it would be appropriate to give clearer and broader definitions of the individual "personality", as well as "Soul" and "Spirit". In the language of spiritual science, "personality" is defined as the three lower bodies of a person - or four if the etheric body is considered separate from physical; the other two are the emotional body (the desire body, astral body) and mental body. We have already talked about the "levels" or "planes" of being, but we need to return to this topic from time to time in order to move on. Of course, we know well what our physical body is, and perhaps we take everything related to its vital activity for granted. In fact, Life is provided by the presence of an etheric or energy body (sometimes called vital, that is, "life"). When our energy body disconnects, it means (physical) death. (In the next section, we will talk in detail about our energy body.)

When we sleep or are unconscious, the connection with the higher bodies is maintained, but they do not necessarily penetrate the physical body. In fact, "life" on the physical plane is decay (and this can be seen when looking at a withered plant or a dead animal), as it breaks down into its constituent parts in order to become something else. Of course, this function is very important at its level, but it plays a secondary role when the body is occupied with Life. In other words, our physical body is nothing more than a suit in which it is convenient for us to receive our lessons, but it is not eternal and when we We wear out the "suit", we must get rid of it in the most hygienic way. This is one of the reasons why cremation is becoming more and more a part of human consciousness and is being resorted to more and more often: cremation purifies and releases energies for new use, otherwise they would gradually decompose and pollute the environment.

Therefore, there is much more sense in cremation than inwasting energy and valuable materials on an already useless corpse. It is very important to understand that how we live now determines how our body will be in the next life (and this is another reason why we should follow the guidance of the Soul). In fact, by our actions we create all future conductors (bodies) for the next incarnation of his personality, including astral and mental. Our emotions and desires are well known to us, but we must also be aware that they exist in a special, vast and potentially dangerous "space" - on the astral plane. The danger is connected with the fact that at its lower levels, in the "astral world", the collective fears, anger and hatred of mankind are hidden - the seeds of violence. Unfortunately, many people spend most of their time on the astral plane. Therefore, it is very important to "calm the waters" of our emotions and develop self-control. And then we will have a clear reflective "surface" on which higher spiritual energies can be imprinted.

The teachers of mankind have always used the symbolism of water when they gave their instructions on the astral (emotional) plane; therefore, by considering the qualities of water (liquid), you can learn a lot about it. When the vibrations of water decrease, it becomes hard and cold (ice); when the vibrations are too high, it turns into steam (transition to higher levels). Water "drop by drop wears away the stone"; it dissolves minerals. In the same way, the higher realms (mental and spiritual) destroy and consume the lower ones (physical and astral).

All our desires and emotions cause the secretion of various fluids: anticipation is associated with the release of sweat or saliva, joy and sadness - with tears, intense fear - with urination, sexual arousal - with the release of the corresponding sexual secrets. When we get sick, our body also releases fluids in different ways and in different places. This connection is unconsciously reflected in our vocabulary: experiencing strong emotions, we "boil", "freeze", "melt", "pour out feelings", etc. We have already said that the universe is our teacher. Therefore, in everything you need to look for conformity!

A great teacher from the East said: "To get rid of suffering, first get rid of desires." Gradually conquering your desires we will definitely feel how our suffering decreases and we become happier. We have already talked (and will continue to talk) about how important it is not to get attached to anything. Now let's move on to the mental body of a person. The lower or concrete mind is that part of our mindwho prefers to take everything apart and analyze. He prides himself on his logic and, as mentioned in the previous section of the book, he is called the "killer of the real" because he does not see the whole picture of the universe. (This is the prerogative of the Soul.)

Delusions of the mind are much more insidious than the illusions of the plane of emotions and desires, and just as exciting. Those people who go through the stage of polarization at the lowest level of the mental plane are convinced that there is nothing but the physical, and that incredibly complex life - and in general the entire manifested universe - arose as a result of a series of random events. Such thinking is based on the belief in such absurdities as: "if a sufficiently large number of monkeys are allowed to play with a typewriter, at least one of them, sooner or later, accidentally "stumbles" on a literary work of genius."

Concrete thinking has led some fairly intelligent people to the delusion that our entire planet, with its amazingly beautiful andcomplex, self-sustaining, self-improving, self-regulating and even self-conscious life appeared by chance, according to the laws of probability!
If I offended any of the readers, I apologize. But such beliefs are the result of limited thinking, and it is time to challenge them. It's time for humanity to wake up; It's time for people to start really thinking, asking and solving hard questions, and not just taking on faith someone else's erroneous assumptions. As mentioned above, the biggest and most dangerous illusion of the concrete mind is the illusion of separation. The higher mind knows that everything United! But we all have to

go our own way in order to free ourselves from the shackles of the astral plane and its emotional "charm".
Even this information is enough to easily understand why our little selves give us, as well as all members of the human kingdom, so many problems.

Human nature is such that we are all focused only on ourselves, we are only interested in "I, me, mine", only our own physical body with its appetites, only our desires, which invariably lead us to a dead end, and our very limited mind, busy in mostly their own illusions. (Now we are not talking about the abstract or higher mind, which is part of our spiritual self.) All the time, throughout many lives, the Soul observes and gives instructions to the personality, which continues to improve, until, finally, it becomes clear that the personality has developed well. The soul knows that now the person has to build a rainbow bridge that will connect the personality with the higher spiritual "I" (which has always existed on its own planes).

But there is one problem here: the personality loves all things as they are; she is satisfied with the situation, she likes to command, and she is not going to cede her power. Interestingly, in the Teachings of Wisdom, the human personality (at this point in evolution) is called the "Guardian of the Threshold": after all, it wants to maintain its control and prevents us from reaching and connecting with our higher, or spiritual, "I". This is the main cause of all human suffering. inferior,the mundane "I" constantly resists the guidance of the Soul to the penetration of its energy. Ultimately, the whole conflict comes down to the resistance of matter to Spirit (and we are still largely matter). Its result is pain, which occurs immediately or later, for "as you sow, so shall you reap" (in some traditions this is called "karma"). It doesn't take much imagination to imagine how the world would change if most people focused not on their own personality, but on your spiritual body. Even now, in the presence of a person whose personality is "impregnated" with the Soul, one feels inner peace, light and a great desire to do good!

So that was a simplified description of the person. And what is the higher, spiritual "I"? Our spiritual triad, or spiritual bodies, exists on the planes (in "spheres") of the three Divine attributes that we have already mentioned: Divine Will, Love-Wisdom, Higher (abstract) Reason. They form the Holy Trinity, or the three Rays of Aspect out of the seven Divine Cosmic Rays of Energy. It is difficult to explain and truly understand, because our spiritual components are still ephemeral because we nourish them too little. But we all have moments sometimes when we rise to the heights of beautiful thoughts, creativity, wisdom, pure love and see a glimpse of our true potential.

Now our planet and solar system is going through a long period of growth, and the most important quality that humanity needs to develop is the quality of the Second Ray - Love. Our God is the God of Love. In the previous life cycle of our solar system, our God was (primarily) the God of mind and activity. This is the sequence of spiritual development: first we gain intelligence, and then Love (and we can love intelligently). Now we have so much intelligence (without love) that every problem imaginable is thrown at us. It is still difficult for us to understand Love on a spiritual level. What we think of as love is mostly love for ourselves or for our fellow human beings. We are just beginning to acquire the quality that the teachers of humanity spoke about: love for those who are far away, love for enemies. Let's dwell on this important point in more detail.

The first thing that comes to mind is: how can I love someone who I don't like or who I don't even know? This is the whole difference between personality and our higher, Divine "I". In passing, we note that at the current stage of human development, our spiritual "I" is represented by the Soul. But in the end, even the Soul will no longer be needed for us - we will ascend into the very kingdom of the Holy Spirit. We have already said that another problem is our modern language. It is easy to understand why so much of the written wisdom of the world is based on ancient languages: they (Sanskrit in particular) have words and expressions that express spiritual realities much more accurately. Holy Scripture translationsmodern Western languages are often corrupted, and we have to borrow words from other languages in order to better express deep truths.

But let's get back to Love and try to understand it. Let's start with words like "compassion" and "sympathy." The highest meaning and the most subtle meaning of such words as "intuition" "pure mind", "understanding", "purity", "integrity", "care", "truth", "sympathy", "courage", "enlightenment", "grace", "favor" will help to better reveal the meaning of true spiritual Love. It's something very far away from a sentimental, selfish, sex-related "love" personality. As soon as we begin to see other people as they really are, that is, beings, like us, passing the path of evolution (consciously or not), their features will become clearer to us. When I see myself and most of humanity as the children on the spiritual path that we really are, it becomes much easier for me to understand others (and myself); then love sprouts for everything and everyone. A higher perspective opens up and you begin to realize what spiritual Love is without any conditions. That's when true compassion begins to form.

Evil

Speaking of Love, one should also mention its absence — that which we call evil. Good and evil are not determined by some arbitrary laws, sent down to us by some incomprehensible deity. Good is what turns out to be the greatest good for most people; evil is that which causes harm and suffering. Everything seems so simple; but we keep hurting ourselves and others.

In terms of spiritual energy, Love and Light are two aspects of deity, and the opposite of Love is fear.
Therefore, when the light of Love is obscured, the shadow of fear appears. If we let the Light in, then fear will turn into Love. If we do not do this and allow the shadow to become darkness, then on the astral plane fear will grow into hatred, and on the physical plane it will turn into violence. A vicious circle sets in: fear breeds hatred - which leads to violence - which breeds fear, and the snowball grows and grows. This is how evil works: everything begins out of fear!

Whenever someone sows fear, all this plays into the hands of the dark forces! This is not about those justified big and small anxieties that are inevitable on our human path. They can be dealt with in a wise, enlightened way. We need to emphasize again: at the level of matter, everything seems to be separate. Matter, on the other hand, has correspondences on the lower levels of all planes (astral, mental, etc.), because these levels, in essence, represent the coarsest and heaviest energies of the corresponding planes. So, when the lower levels of the emotional or mental plane are involved (and they often are), we perceive ourselves as separate from others, and in this case, a shadow of fear easily arises.

Essentially, all evil comes from the illusion of separateness and its echo, the illusion of lack. The universe is abundantbut we humans create our own disadvantage by our greed, ignorance and stupidity. And we begin to believe that we can do something for our own benefit, even if it hurts. harm to others. Having gone through this stage and realizing that we are all part of a great Unity, we really begin to "do to others as we would like them to do to us", because if we are part of God, or the Universe, then others are us and eat! We feel this connection even on a personal level when we move on to higher feelings, such as parenthood or romance. We must understand that at the highest levels we are part of the Universe and are connected with everything that exists in it. At these levels, all components of the planetary Life are interconnected, and it is directly connected with the solar Life, which is an integral part of the Cosmic Life (or God). This explains why Divine Beings identify themselves with All that is, and why the Soul manifests itself in true compassion on a human level.

Sympathy is the lowest correspondence of "Divine Identity!" Once we understand this, and there will be an end to wars, crime, and we will no longer intentionally hurt other people. Then we will truly follow the Golden Rule and begin to treat others the way we want them to treat us. We are one humanity, one planet, one solar system, one cosmos, and all this is part of one Life. Therefore, humanity, when it eventually unites and becomes enlightened, will make the Earth a sacred planet. If we could see the whole picture, to see the full scope of human evolution, to see how we eventually learn the necessary lessons and, growing up, do no more harm to ourselves and others, then evil and suffering would take their proper place in this picture.

Pain and suffering, as we experience them, are temporary conditions! And the birth of a child is usually associated with temporary discomfort, and it is difficult to care for a baby. But, when children grow up, all unpleasant moments are forgotten and communication with them brings joy. We need to understand that we are all "children of God" and, having lived countless lives, we will come out of the initial stage of ignorance; having experienced pain as a result of wrong actions, we will eventually direct our energies to good deeds! As our consciousness grows, we create more positive karma rather than harming ourselves.

Evil prevails in the world mainly due to the thoughts and actions of people on two levels. On one level, the lower astral, we succumb to the inertia of matter, we are seduced by the sensual side of things and material life and we want to have them forever. This is the result of stupidity and ignorance (one might say "sin of omission"). It can be overcome by engaging our higher mind and "will" and doing what we know is right, raising the energy of matter to a higher level, not allowing gross matter to drag us down.

On another level, the lower mental level, there are thought forms created by those who deliberately support the dark forces and try to prevent the enlightenment of people. Here the "sin of allowing" reigns. These energies are fed by those who love power and are seduced by the illusion of the importance of their person. Such people, focused in the lower mentality, are more dangerous. The forces of evil use such people to foment wars, because good people are involuntarily involved in wars, who are forced to kill and destroy, protecting themselves.

What we sow is what we reap. Don't joke with God! Those who obstruct Light and Love, even if only in their thoughts, would immediately stop doing this, if they knew what chain of events they provoke, and that all this will turn against them. After all, the energies of evil can be born even at the subconscious level, and we need to control our thoughts, because they can lead us far. You can often hear the question: if there is a God or higher Beings, then why do they not interfere in what is happening and do not prevent evil? This question itself reflects a lack of understanding of the design and purpose of evolution and the role that we have to play in it.

The eradication of evil is the main task of the human kingdom! We need to remember that matter is (relatively) unenlightened substance. And evil in human dimensions arises from the lack of Love and Light. And, although we are still at that stage that can be called "pre-divine", in, as it were, on the eve of our Divine destiny, first of all, it is we, the people, who play a key role in eradicating evil. Our (human) purpose is to bring Light: it combines with matter and creates all manifestations of Love. Evil is defeated only by Enlightenment!

In other words, we were all created as part of the Divine Plan, and along with all the other components of our universe, we are destined to be co-creators. This is one of the reasons our kingdom exists. How else would we grow if we were never faced with a choice and if someone else did our work for us? We are not here for a walk!

Let us emphasize again: we, the human kingdom, like all other kingdoms, are destined to raise the consciousness of matter; lift and thereby free it, and not allow matter to pull us down and not hold us back.
To do this, it is very important to open your Heart (heart center, or chakra). This is necessary for ourselves - for all mankind - and for all other kingdoms that make up the planetary Life. At some level of our being, we all know that the world as it is usually presented to us is not a reality, and that many of the values of our society are false values! For example, imagine how different the world would be if we honored and cultivated altruism rather than greed.

Note that greed is propagated everywhere openly, aggressively and openly, while altruism is only talked about. What if the models to admire and emulate were altruists, compassionate people who actually do good? But we live in a world where infantile people with the lowest values, who indulge their whims all their lives, are considered "prosperous" just because they have obtained money or temporal power from the system and use it for self-aggrandizement. The day will come when humanity will reach a more mature state on the path of evolution and our society will be wise enough to completely correct this delusion.

In short, human enlightenment is gained through: meditation, which at first can take the form of prayerful contemplation: we become open to the perception of the Higher heavenly influences. Sincere and constant study is the study of higher truths in all their manifestations. Attitude to life as a service for the benefit of the entire planet.

Meditation, study, service – this threefold Path allows us to begin to feel the incredible reality ourselves, in which higher dimensions of existence are open to us! And not only are they open, we are encouraged in every way to enter in order to participate in them and make our contribution. It is interesting that in the higher esoteric teachings it is said that what we perceive as Love is the lower reflection of the Law of Magnetism, the Universal Law, which keeps even the planets and solar systems in its orbits.

At the beginning of the section, we gave examples of how we are drawn to the past; now we are talking about the attraction of the Cosmos; a thinking person has something to think about. So far I have tried to install the following importantprerequisites:

The universe consists of numerous levels, degrees and units of energy, each of which has its own consciousness. All of them are perceived as matter, life and space. At our (human) level of spiritual development, our very life, environment and every experience of life is our teacher. The root of all evil is in attachment to the material and in the illusion of separateness. We are "Soul" and "Personality". The "I" that clings to the past is focused only on itself and stretches down tomatter. The soul, or our adult "I", is directed forward, outward and upward; it takes care of the good of the whole and the growth of consciousness of lower and grosser levels (matter).

In essence, any conflict is a conflict between the Soul and the personality. Therefore, pain mainly arises as a result of friction caused by the personality's resistance to the call of the Soul. What seems to us to be crises in our personal lives are actually manifestations of spiritual crises. All of the above can be considered an introduction to the spiritual life for the sincere seeker.

Energy Centers, Planes, Bodies

Scene: living room. Young woman sitting in a chair and reading a book. The father enters the room.

Father: Hi, how are you? What are you doing?

Daughter: I am reading a wonderful book on chakras.

Father: Again? Listen! You know in your heart that this is all nonsense! Get it all out of your head! These are your gurus, or whatever they are, they are already sitting in my liver. I would give them a kick in the ass! I know, I know what you're going to say. That I'm a narrow-minded materialist.

Curtain.

Here you go again: The Higher Self knows what the personality rejects. Even people who have been led to disbelief in the existence of spiritual bodies and centers of higher energies, in everyday communication unconsciously mention the main (or secondary) chakras. How can this be! Why do we so often choose to remain blind (meaning the "third eye")? Why do we continue sleep when we only need one: wake up and see the truth straight around you? How can you deny? In all languages of the world, the word "heart" is associated with the qualities of pure love, compassion, sympathy, altruism, courage, etc. The qualities that are now being introduced into the consciousness of mankind ("God is Love"). Qualities that humanity so desperately needs! And that's just the heart chakra. What about the other seven (that number again) major energy fields that energize us humans?

But stop. First, it is better to dwell in more detail on the energy (etheric or vital) body, which has already been mentioned. The fact is that energy centers (or chakras) do not exist in the physical matter of our body, but in the energy bodies penetrating it. It should be noted that ethereal matter is in fact physical, but so subtle that mankind does not even have instruments to detect it, except for some part of the electromagnetic spectrum (this includes some ethereal auras that can be captured using a special photographic method, and I believe , what is called
"morphogenetic field").

Since these energy centers do not exist in the physical

body, but in the etheric (and higher) bodies, it must be understood that their names, which refer to the physical organs (heart, throat, solar plexus, etc.), are only approximate indicate their location and relationship with certain bodily functions.

The ethereal substance not only penetrates everywhere, but also connects everything with the All.
Through the ethereal fields, we humans are "connected" to all life on the planet, including Planetary Life itself. And Planetary Life, through this energy, is connected with the solar system and Solar Life. We have already talked about this: thanks to these subtle-energy connectionswe are part of God. Understanding this, it is easier to perceive the universe as a hologram and realize that everything is contained in Everything. Learning about the ethereal or vital energy, about its all-pervasiveness and that it is true life on the physical plane, we begin to better understand the whole universe and realize that what we feel physically is only a shadow of what is real. exists.

We could talk more about this important aspect of reality, but we must return to the main energy centers.
Before we look at the seven main (there are still secondary) centers, it is important to emphasize that in the human body, the diaphragm symbolically separates the four upper, or Spiritual, energy centers from the three lower, or personal ones. It is very important to remember this, because as our consciousness grows, our "lower" energies are transformed and transmitted "supreme". In fact, we are building a bridge, a "rainbow bridge" (called the antahkarana in Sanskrit) between our personality and the Soul, to help this process.

And now let's talk in more detail about the seven main energy centers. Let's list them from top to bottom:

Crown Chakra

The energy field of the crown ("crowning" the head and the whole body) seems to embody the crown of all human achievements on the spiritual path. Through it, as well as through the heart, we are directly connected with the universal Divine Spirit. Depicting awakened beings, spiritually sensitive artists often draw a halo around their heads or a halo above the very top of their heads. Sometimes we unconsciously try to reproduce this crown center on the physical plane, to create its surrogate. That is why, throughout history, the rulers of all countries of the world "crowned" themselves, vainly (and vainly) believing that this adds wisdom and superiority to them. In this sense, those primitive tribes are wiser, in which the applicant for a special headdress, which plays an important role in rituals, must convincingly prove his right to wear it, demonstrating courage and maturity.

Third Eye Chakra

It is the inward-turning eye which, as our consciousness evolves and we come into contact with Soul awakens and becomes the so-called "Ajna center". All knowledge, all information is already "here". In the Teaching this is called a "cloud of knowable things." (See, for example, "Treatise on White Magic", orig. p. 456., referring to Patanjali - apparently, "Yoga Sutras", 4:29). And we can touch this huge storehouse of knowledge (and we do it!) more and more as we become enlightened. At this stage of the evolution of consciousness, this center is still rather poorly developed in most people. But everything changes when we become familiar with the process of visualization and begin to use it to consciously create on the level of ethereal and mental matter. As a result, the chakra the "third eye" begins to act, and we get more and more inspiration.

Mankind is still little aware of the enormous power of inspired (that is, spiritualized) imagination. By activating higher imagination (not to be confused with mere daydreaming), we open ourselves to inspiration. Then we must seize this inspiration, strengthen it and energize it through the developed ability to visualize, and begin the creative process of building thought forms of great potential. Thus we begin to create in a higher reality, as we did before. through our carnal desires - in astral matter. And this is just the beginning. All brilliant creators of the past and present, in whatever area they apply their strength, have something in common: a developed, spiritualized imagination.

What changes next is that as our consciousness grows, the pineal gland and the pituitary gland will gradually begin to interact, as a result of which our latent intuitive abilities will be revealed. How much would humanity change if we used pure reason, or "direct knowledge" (which already exists on the higher planes)! In everythingtimes, enlightened beings have demonstrated this ability. When people's intuition is sufficiently developed, we will no longer be able to deceive each other, as we often do now, because we will see through the lies. It is important not to confuse intuition with "lower psychism." The latter is based on the solar plexus center and focuses mainly on the astral plane. For a developed person, Ajna ("third eye") becomes the "eye of the Soul", its "window to the world".

Throat Chakra

interesting because it is the energy center of our higher creativity. This spiritual center works to one degree or another for all talented people of art: artists, sculptors, architects, musicians, etc. Over time, this center, like all other chakras, will open (or get enough energy) for all of us, if we make the necessary efforts to expand and grow our consciousness. At the same time, the energy of the sacral chakra, or the sex center, which is now used for reproduction (and in fact, more for entertainment), will be transformed and will rise to the throat chakra.

Even from a physiological point of view, there are some correspondences between the throat and the reproductive organs, more precisely, between the tonsils (tonsils) and the sex glands, or gonads. If you think this sounds ridiculous, think of some diseases - mumps, for example - that affect both the tonsils and the testicles or ovaries. Science cannot fully explain the role of the tonsils in the body (I suppose this is a matter for the future). Damage to the seminiferous canals in men directly affects the vocal cords and the voice changes.

Here's another example: I've heard that some mentally handicappedyoung people have exceptional abilities in some area of the arts. But upon reaching the age of puberty, they lose their gift (it is replaced by sexual attraction). Again there is a connection between the sacral and throat forms of creation! Interestingly, animals, unlike humans, are not capable of passionate kisses in sexual relationships. (Not to mention the pleasures of oral sex.)

Chakra Of The Heart

Although we have already said something about the heart center, it is now very important to realize that humanity needs to develop qualities of Love-Wisdom in this, our present, solar system. The reason is this: we are now living in a second ray solar system, and one of its main purposes is to imprint this Divine quality on humanity. This is true, for all the world's religious teachings say that our "God" is the God of Love. Being in the aura, or energy field, of this great Being, we will gradually absorb these spiritual qualities of the Divine Heart (despite the fact that people are very resistant to all new and unknown energies). What a wonderful time it will be when this happens!

One can imagine how our lives would change if peoplebegin to treat each other the way they would like other people to treat them. After all, then antisocial behavior and wars would be simply unthinkable. Perhaps the time has come to note that sometimes the energy nodes in the chakras are compared to lotus petals. When the "petals" of Love open in our heart center, we will become truly loving beings. Already now, many people have their heart centers open, and soon their number will reach a critical mass. It was truly said: "The meek shall inherit the earth" (see Ps. 36:11,

Matt. 5:5).

So far we have talked about the four main energy centers,located above the diaphragm, which are called spiritual centers. Now let's move on to three important centers, which are located below. diaphragm and associated with personality.

Solar Plexus Chakra

In the physical body, the solar plexus is like the "brain" of the viscera. The chakra associated with it governs our emotional life and our desires (but not high aspirations). It is here that the less spiritually developed people are polarized - and such people are still the majority among us. The energy from this center is gradually transformed and rises to the heart center. If someone "swallows" their emotions instead of wisely and with love to understand them, this often causes problems with the stomach or digestion, such as an ulcer. When someone overwhelms us emotionally, we say that we "can't digest" such people. We say about something funny: "the stomach can be torn": laughter is also a reaction of the solar plexus center.

The Sacral Chakra.

We already mentioned it when we talked about the throat chakra. This is the sexual (reproductive) center, which is associated with self-esteem and controlled instincts.

Root Chakra:

this center, located at the base of the spine, is associated with metabolism, with many functions of the body - digestion, blood circulation, excretion, etc., - fromon which our physical health depends. The excretion of gross (solid or liquid) wastes by the corresponding organs can be compared with how gross matter is forced down on all planes (and good energies rise up). The speech of many people who are most focused on their two lower chakras is replete with unconscious references to these centers. The "obscene" words almost exclusively refer to the physical organs corresponding to the lower chakras. The most offensive swear words are related to the genitals or excretory organs. It is interesting to note that it is those who are most "centered" in their lower centers who treat them with the greatest contempt.

It should be noted that there are two chakras (or double chakra) associated with the spleen, and it is also considered an important energy center. (We'll talk about the spleen later.) There is some connection between chakras and planesconsciousness: the heart chakra corresponds to the level of Love-Wisdom (buddhic); coronal correlates with the highest Divine plan; the "third eye" chakra — with the causal plan (the plan of the Soul); the solar plexus and sacral chakras, respectively, with the lower mental and astral. Although all rays affect all chakras to some extent, some chakras resonate more with certain rays at any particular stage of evolution.

And speaking of chakras, the human realm is the only

physical realm that walks and stands upright (some species of birds, which are more oriented toward the deva realm, don't count). The reason is that our higher centers must be placed vertically. This was not until each person was given his own soul (which was the beginning of the human kingdom). In the animal kingdom, the corresponding energy centers are located horizontally, because animals mainly study "horizontal movement". Therefore, they cannot raise their consciousness higher. Our "mobility" is directed upwards, towards the higher consciousness.

This is why we are taught to meditate while sitting upright: this posture symbolically aligns us (in particular, our spine and main energy centers) with our higher self. Higher energies are also located at the base of the spine. This potential energy is called kundalini and is much talked about in spiritual teachings. If we live correctly, in Love and Wisdom, this force naturally rises up and activates our spiritual energy centers in the right sequence and combination. If this process is coordinated with the proper expansion of consciousness, there is nothing to worry about. But it is important to know that you cannot joke with kundalini: it is a powerful force, and if it is released incorrectly, the consequences can be the saddest - up to spontaneous human combustion!

In addition to the vertically located spine (and chakras) and individual Souls, each person has a third unique feature - this is the larynx, thanks to which he can speak. The larynx allows us to express our thoughts, communicate and create in a big way. As already mentioned, sound has a much greater creative (and destructive) power than is now commonly believed. But again I want to remind you of the good (or harm) that we inflict on ourselves, being under the influence of a harmonious sound (or, accordingly, disharmonious). Rough noise is harmful to us, true music is good, whether it is a human creation or the natural sounds of nature.

In the past, people knew much more about the power of this energy, and the use of sound energy allowed them to erect huge structures of stone (many of which have survived to this day), which, even with our current technical capabilities, amaze us. We still have much to learn about ancient civilizations, and then our ideas about their insignificant abilities will vanish like smoke. But, as usual, people misused this knowledge, and knowledge was allowed to be gradually forgotten. We think that sound is noise. But we must remember that there are sound waves that a person cannot hear. The strengths and capabilities of this sector of the energy spectrum are already being used, for example, in medicine.

Sound is something opposite to light (or, perhaps, its lower reflection). Sound travels well through dense matter and cannot travel in a vacuum, while light travels best in "empty" space and does not travel through most solid materials. The fact that some people are sometimes able to see sound or hear colors confirms the existence of some correspondence between these two types of energy. Individual Soul, vertical arrangement of chakras and larynx(a tool of speech) - that's what helped a person to step beyond the animal kingdom and, in the end, to reach the level of civilization and culture (and not at all the outstretched thumb and other alleged physical advantages that scientists talk about).

Now people are becoming more enlightened, and soon we will learn even more about chakras, or energy centers. Even now, when someone or something makes us experience strong feelings, the localization and nature of sensations in the body - in the chest, in the stomach, in the groin - about many things.speak to an understanding person. These are the reactions of our chakras. Be aware of them. And, since we live in an energetic universe, we should think in terms of the ascending, unwinding spiral of life and the law of correspondence. This means that the physical and spiritual growth of people, as well as representatives of other kingdoms, as well as higher beings, depends on energy centers. By understanding this, we begin to realize why and how we are part of God, or the thinking universe.

The human kingdom is not only becoming the physical nervous system of our entire planet. It develops the thing and into the energy center ("throat") of the planetary Life. And the planets (more precisely, their higher "bodies") are the energy centers of solar Life. (Most planets are not "dead." On the contrary, on many of them Life exists at a much higher level than ours.) Solar systems are the energy centers of the constellations as Living Beings - and so on, up to the entire Cosmos (visible and invisible), which is also a Being, called in religions "God". So it turns out that we are actually created "in the image and likeness" of God. Talking about the energy bodyman and his centers, it is worth noting that they have long been known to many world cultures, and they are not only recognized, but also work with them. That is why Eastern medicine, which deals with the energy body, its chakras, meridians and special energy points, cures diseases that are incomprehensible to Western doctors (thinking is limited to the lower levels of the physical plane).

Having gained some understanding of our energy bodies, we can already explain why people sometimes continue to feelamputated parts of the body: because the corresponding part of the vital body is still "in place". Another example: when blood circulation in some part of the body is interrupted and then restored again, we feel painful tingling sensations - this returns our etheric body to its normal state. We twitch in sleep when contact with our vital body is suddenly completely cut off. What we call "shock" or "fainting" occurs when the etheric body separates from the physical body. This is a protective measure so that people (and animals too) are not excessively injured when threatened with death or in severe pain. Losing consciousness or fainting, we may die (or maybe not), but for us it will not be so painful.

In the future, when mankind becomes wiser and gains more knowledgeabout the etheric plane and the vital body, what now seems impossible, will become habitual. It will be possible to restore (re-grow) damaged parts of the body and organs. But we must be realistic: there are good reasons why we (physically) sooner or later do not care "wear out" and "die". As we understand more about the nature of etheric energy fields, we will be able to understand how they work in other realms. We will be able to explain why animals that are better at perceiving energy fields can anticipate earthquakes, migrate over long distances without any prior training, find their way home without error, and sense "spirits" (which are energy fields). The life of the plant kingdom is also closely connected with the ebb and flow of etheric energies, which is why it is so important to plant plants at the right time.

But back to the information about the vital (or ethereal) energy body of a person. Like our other bodies - emotional, mental and spiritual - it is also located on "levels", or "subplanes", of which there are seven in total. On the etheric energy plane the three lower sub-planes (solid, liquid and gaseous) make up what we call "matter". In other words, everything that we perceive as our physical world. The next two sub-planes, located above, are connected with the vital energy that nourishes the organic bodies of all living things. And, finally, two higher bodies make up a sphere that is connected with the energy "from above" (planetary and solar sources) and attracts this energy "down". Many believe that the so-called the "electromagnetic range" is a sub-plane (or sub-planes) of the etheric plane.

At the beginning of its descent, the light from the Sun penetrates through the (upper) etheric levels as a wave, descending into the grosser levels, it becomes subatomic particles, then atoms, then, when the atoms combine into molecules, what is considered to be matter is formed. On theAt each stage, the light becomes "heavier" and loses its freedom. Then the inert molecule begins its ascent through the realms of nature (cells, organs, plants, animals, people, etc.), regaining more and more of its freedom, and eventually becomes a free being of Light again. From Sun to Soul! The most subtle "matter", or energy, of each of our energy "body" rises to its higher sub-plane, where its essence is abstracted into a permanent "memory", or record of these energy bodies, in the so-called "Permanent Atom". The Permanent Atoms of all our bodies are located on the higher subplanes and remain with us for many lifetimes. These are the "seeds" or higher correspondences of our genes, and "bodies" are built on their basis in each new incarnation. It should be very interesting for all of us to learn how to gain, maintain and strengthen the health of the vital (etheric) body while we are in the physical body, because the etheric body is the health body of our physical organism.

Many people in the so-called (and needlessly) developed world are in poor health and suffering from illnesses because we don't realize how important it is to be aware of these energies and understand how they affect us. Not only fresh air, sun exposure, exercise, proper nutrition (especially fruits, vegetables, cereals, nuts, etc.) have a beneficial effect on our energy body. Since all our bodies are in fact energetic, our thoughts, feelings, and actions also have an impact. And the larger energy fields we live in—physical, mental, and emotional—also affect us, for good or bad. People have often noticed that inner health and beauty contribute to *external* health and beauty. The reverse is, of course, just as true.

Life energy (also called "prana") enters the human body to a large extent through the spleen and the energy field associated with it. As we grow spiritually (our consciousness grows), all of our energy bodies will connect us to their respective higher sub-planes or realms, and our true power will increase proportionately. Of course, this is only a general and very simplified picture. What is especially important:our body needs to be cleansed periodically, and we should welcome these cleansings, take them for granted and not try to suppress the physical discomfort. Listen to your body and act with it. Don't fight it - it will only make the problem worse. The time will come when the present will appear in our society. "health", and then we will begin to find integrity again.

Ritual can also play an important role in the health of our vital body. That is why the higher Beings imprinted prayers, hymns and other ceremonies on our religious consciousness. Therefore, in the West nowmore and more engaged in meditation, recitation of mantras and the practice of yoga. If done correctly, this is to the benefit of our higher bodies. When our physical body is injured, the imprint remains in the etheric body penetrating it. Therefore, scars, wrinkles, etc.remain, although the cells of our body are constantly renewed. Birthmarks (and even some "birth defects") are often associated with severe physical damage suffered in a past life. They are imprinted on our vital body and carried by our etheric permanent atom, which remains with us for many incarnations on Earth (although the "defects" are usually "healed" in one or more lifetimes). Often accompanying (or independent of them) psychological traumas will be imprinted in the form of fears (phobias) in our astral permanent atom and are transferred to the next life.

All planes - astral, mental and spiritual - contain a permanent record of Life and all events. Our "Solar Angel" and other Higher Beings have access to these "chronicles". Speaking of scars and wrinkles, if we accept that fingerprints are unique, and scientists believe that they can determine a predisposition to certain diseases, then why do many deny that palm lines, which we are born with and which are also unique, can do anything? then mean? Think about it: why would a newborn baby have wrinkles on their hands? Palm lines can tell us something about ourselves. There are reasons for everything.

As we open to the Light, we begin to understand that everything is part of the interconnected energy of the greater Life. The lines of the hand, the shape of the head and much more in our appearance, like the astrological natal chart, can say a lot to an understanding person. By examining what lies behind these energy patterns, we find that many and varied clues are available to us to help us understand the meaning of life. If you want to know about color correspondences, then the range of ethereal subplanes ranges from pale lilac to dark violet (almost to ultraviolet). Interestingly, violet is associated with the Seventh Ray of Organization and Ritual (Rhythm). This Ray of energy is now beginning to make its impact on humanity, and the resonance between the energies of the Seventh Ray and the etheric energies will open new possibilities for enhancing the vitality of our etheric body.

Over the past hundred years, seventh ray exposure has made many discoveries in connection with electricity. But this is not comparable to what is to be (and fairly soon) to learn about what we call electricity and electromagnetic energies. Ultimately, everything is made up of aspects of this energy (electricity). Speaking of our

energy bodies, one phenomenon should be touched upon, about whichargue and which is sometimes misunderstood: we are talking about racial bodies. As already mentioned, as consciousness develops, the physical a person's "vehicle" or container that contains consciousness,also improved; raising and expanding our consciousness, we are constantly building and improving our "guides". As for our "higher" vehicles (bodies), we build them from a higher "substance" - from desires, from a mental or spiritual substance. Remember that these bodies, like the realms they inhabit, are even more real and longer-lasting than the physical ones.
But let's talk about the physical now.

First, let's once again imagine the whole picture: in fact, we are the Spirit that has descended and partly "encased" in a body of coarser energy, that is, as it is commonly called, matter. It is more correct to say that the point of higher (or spiritual) consciousness is enclosed in the body of lower (material) consciousness. Let us repeat what was said in the previous section: our Spirit arose as a "spark of God", or our highest Monadic essence, or Life. This ray of divinity descended, penetrating into ever denser substance (and into the corresponding spheres), until it reached the densest substance - matter. In turn, for billions of years, this part of matter stretched upwards and, having passed through the kingdoms of minerals, plants and animals, finally connected with the representative of the Spirit, that is, with what we call "Soul".

And so the man was born!

For all its importance, this is just one step in an endless

process. It is important to understand that race, nationality, gender and living in us "spark of the deity" is essentially different things: one is mortal, transient, and the other is eternal. In some traditions, they are represented by a demon (earthly being) and an angel (a heavenly being) sitting on our shoulders. The interaction of our higher Spirit with the lower "matter" of the conductors of our personality gives rise to the third - a sense of self, awareness, the idea of "I Am". We all experience and express it. Back to races: it is known that science has defined them mainly by physical parameters. Spiritual Science, as always, digs much deeper. We are living in the fifth of the seven (that number again) root races in this human wave of life, and each root race is made up of (guess how many) sub-races.

The first two root races did not descend completely to the level of matter, and therefore did not leave any physical traces. The Third Root Race was the first race to exist in physical bodies and to be taught on the physical plane. The root chakra was the main one at that time. But even then, with the first glimpses of the Light, the germ of an individualized thinking being appeared, and humanity began!
T
he people of the fourth race were more polarized in the astral body, or body of desire, they gradually developed the ability to think emotionally and with it the ability to express their thoughts through speech. At that time, the sacral and solar plexus chakras developed. It can be said that they have developed too much, because people sometimes fell into sexual excesses and other vices that surpassed even

current. Because of these degenerate tendencies, most of our ancestors from the fourth root race were finally destroyed inseries of cataclysms. This is told in the myths and scriptures of all the world's cultures, although they were simplified for people of past times. There is also a lot of physical evidence of a global flood, although many of them are yet to be discovered in the future.

The main achievement of the fifth (current) root race is the further development of the concrete mind. Again, somewhat redundant development, with an emphasis on technology, science and logical thinking. Although this phase is important and necessary in the evolution of human consciousness, it is only one step on the endless ladder of the cosmic hierarchy of enlightenment, and even one of the first steps, but, of course, not the main and not the last, as some people think. But even those who are focused in a particular mind will move on to higher levels when this stage has done the necessary work.

We have a much more glorious destiny worthy of the most ardent aspirations. What in esoteric science are called "sub-races" of root races (and "branches" of sub-races) are, in some cases, anthropological "races". To avoid misunderstandings that have already caused great suffering in the world, it is important to highlight the following points: First, when natural science speaks of races, what is generally meant is the physical body, and not the Soul, as has already been said.

Secondly, all races are genetically descended from previous races (with some help from above, which we will talk about shortly). Therefore, there are no absolutely new or pure races. Hence, there is no physical or spiritual reason why people of different races cannot marry and have children. But there are many different reasons why people can do this, and one of the most important is to provide genetic material for new races.

Thirdly, there are no "bad" or "good" races. From time to time, new racial bodies appear that provide the Soul with more suitable and refined vehicles for learning the next lessons destined for us, and the old, coarser "forms" die off. There are many examples in anthropology. In addition, new racial bodies are created taking into account the changing climate of the Earth. Since everything that makes up planetary Life is constantly being improved, and the planet "accelerates", that is, raises its vibration (its consciousness),it is not only the physical bodies of men that change, it inevitably happens in all the kingdoms of nature.

We know that in the distant past, animal bodies were much cruder,and with the advent of other, more suitable vehicles, the former bodies gradually disappeared. Scientists are trying to find the reason for the extinction of dinosaurs. In fact, the dinosaurs were "killed" by the fact that their bodies stoppedmeet new opportunities for improvement. Their wave of life has passed into new, smaller but more efficient bodies. The same thing has happened to many other animal species (and will eventually happen to humans as well).

Fourth, any reasonable person should understand that each race has something to learn from other races.

It's time to talk about racism. Basically, it is born from low self-esteem, which translates into a desire to find someone to look down on. It is known that well-adjusted people with healthy self-esteem are not found among the supporters of extremists and do not suffer from paranoia. Life is a mirror: those who slander other people expose their own weaknesses. Weaknesses that we do not want to notice in ourselves, we project onto others - be it laziness, thieving, deceit, sexual promiscuity or other "sins".

And now we come to the present moment. What about the coming races? To answer this question, we must deviate a little from the topic and recall the kingdom that I have already mentioned and which is called
the "realm of the devas" or angels. This vast and omnipresent realm is associated with many misunderstandings among people. I will try to give my own, extremely limited (and probably somewhat erroneous) interpretation of this important line.
evolution. This realm, which is usually not perceived by the five senses of a person (because its representatives dwell in more subtle realms), has been spoken of by many mystics, psychics and spiritual teachers throughout history, and its inhabitants are mentioned in religious scriptures around the world.
Myths and legends speak of some of these beings, the least developed and most varied, the nature spirits or elementals. More developed beings are often called angels.

At the current level of human evolution, the deva realm and the human realm are considered parallel worlds in a certain sense, although in the process of evolution the devas must also pass through the stage of the human realm in order to reach higher spiritual levels. Therefore, our consciousness and theirs are not fully compatible until we advance into the higher spiritual realms. However, in both realms there are aspects that are deeply intertwined.

Since the evolutionary lifestreams of devas and humans follow a parallel course, they have, to some extent, the same levels of achievement: what we call physical, astral, mental and spiritual. The Devic beings both constitute the matter of these planes and are their builders. In other words, they build from their own substance. This is easier to understand if you think of them as energy. who they are, not how about the forms they create. The lower or involutionary devas who dwell on the planes corresponding to our physical and astral (and even lower) are often, as already mentioned, grouped under the "elemental" group. The imagination immediately draws us witches in pointed hats with black cats and boiling cauldrons, but although people sometimes (at great risk) try to influence these entities from evil or selfish motives, elementals do not have such free will as people do. But they are happy to work, obeying their own high mentors and spiritual mentors of our planetary evolution. (Remember: "Teacher of angels and people"?)

The deva kingdom is especially active in the plant kingdom. The spirits of nature, which are so much talked about, are not the fruit of someone's imagination. They are responsible for progress and growth in this realm (and embody it). Each element - fire, water, wind, etc. - has its own spirit. These elementals have no intelligence in our sense, but they can be quite playful. Has it ever happened to you: you are sitting by the fire, and the smoke is reaching for you, regardless of the direction of the wind? You change seats - he will follow you ... The esoteric teachings say that insects and birds are closely associated with this kingdom and in some cases act as intermediaries between the two evolutionary streams - devas and people. (It is curious that many of the "signs" are associated with birds. Also remember the Holy Spirit in the form of a dove.)

What does all this have to do with the racial body of man? As I have already said, new races are periodically introduced in order to provide more perfect vehicles for our growing consciousness. Some of the unusual phenomena that are happening now may have a direct bearing on this.

Ufo And Devas

We have all heard many times about unusual phenomena that occur almost daily. Although they are often attested and documented in detail, most people have no way of believing them. I mean the well-known UFO phenomenon. Of those few who are not averse to at least getting acquainted with the evidence, most are convinced that these are the tricks of beings from other planets, which are very far from us. It is interesting to note that this category of people can be roughly divided into two groups: some believe that alien beings have good intentions and want to savemankind from ignorance and self-destruction, while others see more sinister and selfish motives in their visits. We again project our own nature and our own fears onto others. But I would like to make a different suggestion. Namely, these phenomena "handiwork" of the devas. Now the realm of the devas, or angels, is helping to develop new racial bodies for humanity (as has helped throughout our history). In addition, they have other missions related to evolution.

To begin with, as orthodox science has established, minor changes and improvements occur under the influence of "natural" genetic mutations. The ability to gradually improve the physical and other bodies as consciousness grew was from the very beginning "programmed" into any life. But is it not possible to admit that for essential changes, which the Divine guides of the human race periodically recognize as necessary, the help of "outsiders" is required? In some religious traditions, the inhabitants of this kingdom parallel to us are called "angels". But, in the end, this kingdom includes both the builders and the very substance of our physical shells. Isn't it logical that it should also participate in genetic (program) changes?

Orthodox science finds it difficult to explain the rapid growth of civilization and culture in the current geological era. Her theories cannot substantiate evolutionary leaps in the development of mankind, and one has to resort to hypothetical "lost links". "New and improved" human models always appear "suddenly", relatively unexpectedly. And so it is not only with the human races, but also with the plant and animal kingdoms: "suddenly" new species appear, and the old ones constantly die out. In times of great change (as now), when the new zodiacal energies coincide with the new combinations of energies of the Cosmic Rays (both of which greatly influence planetary life), it is precisely to expect the emergence of new forms of life. And if so, then why not assume that the famous phenomena of "crop circles" in the plant kingdom, "cattle mutilations" (and in fact, surgical intervention incomprehensible to us) in the animal kingdom and "genetic experiments on UFO captives" in the human kingdom - are these just individual manifestations of the numerous physical transformations that accompany the current psychological and spiritual changes?

It has already been said that the five senses of man usually cannot perceive the realm of the devas. But the converse is not true: in general, the devas know about us. And some of them, under certain circumstances, can even slow down their vibrations and move into our dimension. They can also raise our vibrations so that we can overcome our physical limitations. In this way we can interact in a kind of ethereal "border zone".

It is interesting to note that the participants in the "genetic experiments" associated with UFOs, although they may not want it, find themselves in altered states of consciousness: their consciousness passes through walls, etc. (In another dimension, this is, in fact, a normal state.) Here is another curious detail: they say that the structure of their body and especially the eyes of the "aliens" resemble insects. Such outward forms are easier for the devas to take on than the more complex ones—say, human—because insects and birds have a closer connection with the devic realm. Now let's talk about why these "contacts" with UFOs are perceived as violence.

Imagine yourself in the place of a person who had to endure such a traumatic experience (especially if a person does not understand the evolutionary background of this). And when you try to talk about your experiences, they tell you that either you were misled, or you invented everything yourself, or - if they do believe it - you fell victim to terrible creatures from another planet. Naturally, you will remember your experience with double horror and disgust. But let's look at all this from a different point of view: if we humans are in some sense "cells" of the physical body of God, andour physical bodies change (since we incarnate in thousands of bodies over billions of years), which corresponds to the change of cells in the body of God, then perhaps we should not be so completely identified with our bodies? Instead, we should understand that they are like clothes that we put on in the morning and take off at night, and that our bodies do not even belong to us: they are given to us for temporary use. And if so, don't we want bodies to be constantly improved? This process can and will provide us with better and more appropriate shells as our consciousness grows. After all we have a higher purpose than just to exist.

If we believe the numerous stories of "abducted by aliens" (discarding obvious fabrications) about the experiments carried out on them and look at all this in the above context, will we not see more common sense in these events? And, most importantly, will they not turn out to have more common sense than existing theories? In other words: how else can large-scale evolutionary advances be carried out? Although most people have an idea of angels and devas from traditional religious teachings, we must remember that these concepts are mostly explained to us in childhood; accordingly, this information is mainly designed for the perception of a child's immature mind, and much more is added "for the red word." Therefore, it is important to emphasize that other kingdoms do not exist at all in order to satisfy our fantasies and desires. They, like us, have their duties and their place in the general scheme of evolution (their own dharma, as they say in India). They have no intention of harming us. In a broad panorama, they are of great help to humanity.

But there are creatures both humans and non-humans who, out of ignorance or malice, try to interfere with their work for the benefit of evolution. It follows that in learning more about the deva realm and its role in the Divine Plan, we need to understand that the events in which they are involved are not always simple and can be risky. Therefore, we must be careful not to intentionally interfere with the work of the devas in any case and not to try to use them for selfish purposes. Attempting to manipulate beings from the realm of the devas is what is called black magic - an extremely dangerous occupation! But there are people who can communicate with nature spirits with care and respect, and, moved by love and not by selfishness, they can receive instruction from the devic energies in the vegetable kingdom and cooperate to some extent with them.

When a new universe appears - after a long "night" of rest - it begins with a sound manifestation of matter (or lower Spirit), followed by "Light" (or higher Spirit), gradually deeper and penetrating deeper into matter. This results in consciousness being created at every level (in a sphere or realm); it descends, and thus begins the process of Life. The All then begins the long journey of returning to perfection (or the "House of the Father"; see John 14:2). Countless universes - with countless galaxies - with countless solar systems that unite countless, increasingly complex lives, and all this is forever moving along the ascending spiral of the shining pinnacle of Life! And all this time usliving on a tiny planet, Divine Teachers teach the mysteries of energy at all levels and how to use it correctly in this theater of being. Gradually, we fulfill our role, enlightening our share of the darkness, and thereby taking on the responsibility of enlightening it further and further. Until there is no darkness at all!

Thus, after billions of years, everything comes to perfect balance, to perfect harmony, to a dazzling climax.
And all this is contained in the perfect Cosmic Mind.

School Is Over

Helpless, I sit on a chair nearby, tears rolling down my cheeks. Life is slowly leaving her, and I am in complete despair that there is nothing I can do to help. She is no longer young, but this beautiful woman is still soI could give a lot to this world. How unfair that life ends right now, when its qualities are so needed! Talented, compassionate, self-sacrificing - there are so few people like that! She would still live and live ...

Stealthily I wipe my tears, although who would be ashamed of? It is clear that everyone in this room is experiencing the same feelings as me. If only we could do something! But nothing can be doneand the curtain over her life is slowly lowering. That is life. This is "death". Only death does not happen! Esoteric teachings say that we are born on the physical plane according to the Law of Limitation, and we "die" according to the Law of Liberation. Very soon we will return to what is said in the Teachings of Wisdom about our Return Home. But first, imagine that we are in a theater. Although we know that the actors are acting on stage, the action looks very believable and we experience real feelings. But the performance ends, and we remember that an even more real life awaits us, our real world. Compared to the world of the spectacle, our world has more dimensions; it is still much more interesting to live in it than in the theater, no matter how exciting the production may be. How much more real, interesting and lively our life will be when we return from the theater of the physical plane to our true Home, where there are even more dimensions!

Let's now see what our establishment has to say about this. We are not offered a large selection. One can accept the dogma of modern science that death completely destroys the personality. Or you can accept one of the religious teachings about life after death: either an endless church service awaits you, or eternal torment, the most terrible that a person can invent. It is not surprising that with such a perspective, many people cling fiercely to life. (Interestingly, those who consider themselves the most devout often value life on the physical plane even more than those who call themselves atheists.) We must raise our consciousness and not be limited by these dogmas! We can take advantage of one of the many gifts that are now given to humanity - the opportunity to deeply understand the transition that we mistakenly think of as "death."

Something can be learned from the so-called "near-death experience" (NDE). Such cases are widely described and generally recognized. What answers do they give to eternal questions about death: What does a person feel when the soul leaves the body? What does a person experience when parting with everything he is used to? And what happens after we make the transition? I will present my own understanding, based on the analysis of information available to mankind about the "other side". All those who have experienced clinical death say that they experienced a joyful state. Once they "crossed over" and saw the Light (with the help of the beings that inhabit those realms), they experienced such bliss that they didn't want to go back. Where is the fear?

The Eternal Wisdom Teachings confirm these impressions of NDE survivors and talk about the great sense of liberation we experience when we are no longer burdened by the body that has been limiting us so much. Behind this feeling of freedom comes the realization of wide opportunities to advance towards the Light and thereby strengthen one's spiritual growth. Some might say, well, what good is that? "Spiritual growth" doesn't sound very exciting compared to the joys of the physical plane. But what about the fun? And the parties? What about adventures? What about sensual pleasures? Yes, indeed, "matter" gives us temporary joys (however, severe pain), and it is the seduction of these gross energies that tempts us to return to the physical world, incarnating again and again, until, finally, we outgrow it.

In exceptional cases, the astral bodies of those who are too sensually absorbed may even become "earthbound" after leaving the physical body. Resisting the call of higher life, the remnants of astral energies are clothed in ethereal substance and turn into "spirits". Sometimes they even try to take over the body of a living person. Obviously, if a person is immersed in the sensations of the physical plane and the desire of the astral, he is not yet ready for the deep and eternal joys of a higher and broader life. To give an analogy: if you ask a child to choose between ice cream and going to the theater or concert, most children will choose ice cream. But an adult who is more intellectually developed is much more likely to prefer a cultural event. Since most of humanity is still at the child stage of consciousness development, it is not surprising that we still choose to return to a carefree and frivolous life.

And so it will be until we finally learn all the necessary lessons that are prepared for us on the physical plane. That's when we "Let's put toys aside" forever.

Now that the planet is becoming more and more enlightened, many people will take the opportunity to grow up and choose Life over life. All of the above gives enough reason for relatives not to "keep" the person leaving them. After all, it is obvious that, greatly mourning our departed, we do not provide them with a favorable energy field. Wouldn't it be better to escort them to a new huge world with joy and good parting words? We must also understand that the death of the physical body and brain is a great boon, especially for the human kingdom. Can you imagine how slowly we would develop if we lived forever? Even in the "breaks" between incarnations, many still yearn for the familiar, and in the next life, having new opportunities, they use their free will to return to the old. Another great blessing: we are not given to know our future. What we need to know we get in dreams, visions and signs, but we are allowed to determine our own destiny through free will.

Let's continue talking about our transition. According to NDE survivors, we experience the feeling that our entire past life "passes before the eyes." There is nothing impossible in this, as it may seem at first glance, because our understanding of time is based on the concept developed by our physical brain, which perceives it as linear, uniform and unidirectional. As we leave the physical world and find our home in the higher (finer) realms, we will experience "time" in a very different way. This is what happens in the state of consciousness called "sleep": we dream very a long sleep, and when we look at the clock, it turns out that we took a nap just a little. It also happens the other way around: it seems to us that we have slept a little, but when we wake up, we find that we have slept for many hours.

Sleep and dreams can teach us a lot about what we call death.

In the described process, it is important that we review our lives, re-experience our relationships with other people at all levels. In those moments, we experience happiness or pain - feelings that arose in those with whom we communicated. We face all the joys and sorrows that we ourselves caused, and, accordingly, we feel the same that other people once experienced with us - nothing escapes, no secrets remain. Everything will be remembered - physical pains, emotional experiences, mental torments and all good things. And also the good, the bad, and the ugly.

Since time feels different in this state, sometimes we kind of look at our lives "back to front", and then it is easier to see the causes of many events. This process is somewhat reminiscent of the dogma of purgatory. (Therefore, by the way, the Teaching of Wisdom recommends that before going to sleep we remember the day we lived and try to mentally correct everything that we have done.) You may ask: what about those who serve evil, dark forces? What happens to those beings who cling to the material, who would rather stay in the sensual realm, consciously declare war on any form of enlightenment and Love? What about those who are responsible for drawing spiritually weak people into endless wars, for inciting hatred, fueling greed, for exploitation? Since their energies resonate with the lowest, dirtiest levels of the astral plane, they go there after death. This is a sphere of darkness in every sense of the word, a dimension in which there is absolutely no goodness, truth, beauty. (We humans help create these lower realms with our grossestthoughts and actions while we are still in the flesh.)

This lower level of afterlife would seem like hell to any awakened person. Only beings who have absolutely no connection with their own Soul can get into such an environment. But such people really exist, they are easy to find on the pages of history, and sometimes among us. Some even make their way to power, and they are not only in government, but also in business and even in religion - wherever the goal of division and stagnation can be served. Suffice it to say that we will ascend (or be drawn) to sucha level that resonates with our actions in life on the physical plane and, moreover, gives us the maximum opportunity to learn all the necessary lessons. Everything is there - from beautiful bliss to terrible hells. Indeed,there are "many mansions" (see John 14:2). People who have dedicated their lives to planetary service, have learned to evaluate their actions on an ongoing basis and correct them properly, require only a little experience of being onlower (astral) level, and they quickly move to higher spheres, closer to the Soul. For them, the time spent in "purgatory" passes quickly.

Then we make the transition to the spheres, which in different world religions are called "heaven", "paradise", devachan, etc. During our temporary sojourn in heaven, we are provided with higher opportunities and experiences. There we can further develop the positive qualities that we acquired in previous lives. In the "heavenly" world, we are no longer burdened by the energies of gross desires and emotions - they were erased during our stay in the astral world. Now we are separated from the dark forces.

We can use everything that at a higher level corresponds to human libraries, museums, universities. The higher mental spheres and still higher contain all the most valuable world knowledge and the best of culture.

The time allotted to us will pass (although time is not linear there, but itthere is still!) stay in a higher world, and our unfulfilled desires, karma and needs of the Planet will attract us to a new life on Earth. And then we again descend into the astral plane and again adapt to the energies of this world, because soon we will have a new incarnation and we will be subject to their influence. When the time comes for "reincarnation" (new incarnation), our Soul and the "Lords of Karma" choose the energies of the environment and family (from what is) that are most suitable for the next stage of our growth. I must say that due to ignorance, evil, overpopulation, many of those who return to our world have very gloomy prospects. However, we are given a situation (environment) - again, from what is available at that time - that will provide the best opportunities. It depends on us whether we can fully enjoy each of our lives.

If we talk about further enlightenment, then only a few out of a huge number of people achieve something in each life, because basically a person spends his next life repeating the path he has traveled, he re-learns what he has already begun to comprehend in past lives. Therefore, it takes a lot of time to, so to speak, "pick up speed". And there, our heads are usually already filled with ideas of separateness, because the dark forces want our minds to remain closed. Many people spend most of their lives satisfying material needs and miserable whims, and this is where they see the meaning of life. Therefore, many of us have to live many lives before we finally embark on the path of ascent to the spirit and consciousness, and for this we need a lot of life experience. In different lifetimes, we may be given different personality traits, determined by a particular Beam; we are born under different signs of the zodiac, in different nationalities, and so on. We are given the bodies best suited for the next course of lessons. Gender also changes periodically, so in some life there may be a "failure" of sexual orientation, but over time, both in an individual and in the world, everything harmonizes.

When we understand that a person has many lives, it is easy to understand that why the children of some parents are so different: one child is calm, and the other is noisy, cheerful or cocky. Genetic traits received from parents only contribute to the physical body. The basis of personality has been formed over an infinite number of lifetimes (and will continue to be formed). But personality is also transitory. The Primordial Being is transferred from one life to another by the immortal Soul. It is important to remember one more truth: we have many lives, and sooner or later we will experience (or at least see firsthand) almost the entire human experience. Each of our actions - good or bad - provides for a response (karma). Therefore, for all the lives that we have lived and still live, we, apparently, will cause others, and we ourselves will experience everything that can be caused and experienced. Since many of our deeds were and are bad, they come back to us (karma!) and respond with very unpleasant experiences. But in later lives, when we are tempted to repeat the same mistakes, on some level we will remember how much pain they have already caused us and others.

This is how we begin to develop discernment that leads to wisdom. This is one of the reasons why a "young soul" and an "old soul" find themselves in the same situation and make different decisions. one is incorrect and the other is correct. Of course, "positive" karma is accumulated by the right actions. The Universe teaches us with such methods, and in the end we will learn how to act correctly. I think that when we make the transition and a wider perspective opens up for us, we will look back and life will seem like a normal day at school, of which there are many: the bell rings - and we are glad for a short break. Here I would like to point out that there is much to be learned by thinking about this school model. It is very important to know that this model, which has become so widespread lately, quite adequately reflects Life, albeit at a lower level (again, the Law of Correspondence). And universal free and public education is a very significant achievement in the spiritual growth of the human kingdom. Therefore, the dark forces are trying in every possible way to interfere with this institution. All attempts to make people remain ignorant and limited in their views and beliefs are doing a favor to the dark forces! In order to expand consciousness and grow spiritually, we need continuous study, and it should be encouraged by all means.

Comparing life with a school day, we can continue the analogy: after spending many days (lives) in school, we move to the next class, or to a higher level. We receive a promotion, or spiritual "initiation" (initiation). Although all people (in the big picture of Life) have the same opportunities to advance on the path of Love and Light, it is easy to see that people are at different levels in the school of life. We see that the majority of people are still, as it were, in the "primary grades". There are several reasons for this: not everyone entered the human realm as individuals at the same time (as mentioned earlier). Therefore, those who have been "going to school" for longer, and therefore gained more life experience of life (and experience of lives), are considered "old souls", and they may be a step or two ahead. Another very important factor is that some people put in more effort and use more opportunities, so (as in any school class) they progress faster. And others do not care about studying, they do not see their capabilities and fall behind. Let us emphasize again: it is very important to help each other. it's for everyone's benefit!

Through life experience (study) we gofrom ignorance to knowledge. When the heart chakra opens, we combine knowledge with love and discernment. That is when we begin to gain wisdom. In the teachings, this is called the transition from the "Palace of Ignorance" to "Palace of Learning" and "Palace of Wisdom" (see, for example:

Alice Bailey, "Initiation Human and Solar", p. orig. ten). Here I would like back to the "new group of world servers" that I mentioned in passing earlier. It is at this stage that we stop intentionally hurting others and begin to consciously help others. This is where the sense of responsibility begins. It is at this stage that we become people of good will, not trying to "win over" others, but striving to ensure that everyone wins. Then we have to go through the probationary part of the Path of Discipleship. The soul calls us more and more to serve people, and therefore all Life on the planet, of which we are a part. There are also changes in our beliefs, as we discussed in the previous section of the book. The time of thoughts and searches comes, and when we become open and begin to perceive new ideas, the old ideology no longer satisfies us.

This stage is called "candidate": we strive for spiritual growth, but we still lack the ability to discern. Be careful: it is easy to get carried away with new teachings that sound beautiful and impressive (but may be empty), it is also possible to disbelieve in old beliefs and "throw the baby out with the water." Keep all the best, true and beautiful of the old traditions. And learn to discern. In the end, we stop being amateurs and realize that spiritual work is serious, albeit joyful, work.
Over time, the physical plane and its illusions no longer exert their influence on us, and we begin to overcome the attraction of matter. We begin to focus on higher levels and control our physical desires. This first step is very significant and important. It is much more difficult then to learn not to succumb to the spell of the astral and the world and to establish control over lower desires and emotions. To do this, you need to become

more reasonable, and then the Light will appear, which will dispel the mists of the astral plane. This is the second important step.

Then, when the lower mind has done its work, it, too, must cast aside the illusions of superiority and give way to the higher Light of the Soul, which connects us with our Spiritual Triad (which, I remind you, consists of the abstract or Higher Mind, the Love-Wisdom heart chakra and our Divine Will). This is the third very important stage in ourevolution! Our successful completion of these three (and other) "high school" grades are stages of "spiritual initiation." It has already been said that in countless incarnations our consciousness grows until we are finally ready to "put away our toys" forever and begin to appreciate the Real. Having reached this important point in our spiritual evolution, we finally learn all the necessary lessons of the physical plane, and we no longer need to return there. When most people eventually complete their earthly learning experience, we will become Spiritual Beings. And some "graduates" will take on the role of teachers. Because we cannot see such teachers with the physical eye, many deny their existence. But, becoming wiser, we feel their help more and more. And they are becoming more and more real to us.

The teachers of the school of life are those who help people, and we have already talked about this. In the spiritual traditions of the world, they are called differently:Brotherhood of Light, Spiritual Hierarchy, Mentors, Masters, etc. They are led by the Great Teacher (Savior, Avatar) of mankind. In different religions, he has his own names (titles), but he is recognized by all spiritual traditions. But even in the higher spheres, we will still have something to strive for and something to work for. We will always have access to a new expansion of Life until that distant day comes when the Cosmos becomes perfect and complete. The main content of the book has already been stated, but one more secret must be said. In our time, humanity has to learn another type of energy. The most suitable word for it in our language is synthesis. In the Teachings of Wisdom this momentous event is described as "the coming of the Avatar of Synthesis" (see, for example: Alice Bailey, "The Externalization of the Hierarchy", p.orig. 285-312), and the energy of this great Cosmic Being begins to be felt on our planet. This is one of the reasons people and ideas are now coming together to make the planet what it is meant to be.

We have no idea how great the impact of this energy will be on humanity and on all forms of life on Earth.
We know fashionably: this will lead to a beneficial growth of consciousness all components of planetary Life. Those who have read the previous sections of this book will likely either

a) agree with much of what has been said

b) will consider that all this is by and large nonsense.

One way or another, I am fully aware that only time can confirm or refute the view of the Cosmos presented here. But you will find, I am sure, that your life and your experience do not contradict any of the statements I have made. On the contrary: with them it is possible not only to link everything that happens, but also to substantiate it much better than from other positions. We simply no longer need to try to fit large round rods into small square slots. And for those of you who are ready to stop trying to squeeze your reality into limited belief systems, let me remember: cosmology "mystery schools" were never intended to replace existing creeds or scientific theories. This Teaching is called upon to give people a "big Truth" in which the highest and purest of these worldviews can unite. Basics These views have not been given to mankind in vain, and much is yet to come.

Looking Back From The Future

Let's now look back from our future to the first couple of decades of the twenty-first century and the preceding twentieth century. You can even capture another couple of centuries of the past millennium, when we first began to feel the influence of the coming New Age. There we see a wonderful time of great discoveries and significant changes that occur only at the end of one era and the beginning of another. This is a time of fundamental transformation of the entire planet. Yet we are more interested in the twentieth century. We see in it the Armageddon predicted in world scriptures and myths. A protracted war in three stages.

The first stage was mostly physical - nakedaggressive aggression. The second stage, even more physical, nevertheless affected the lower astral: the ideologies of evil tried to suppress the growing desire for freedom and good will throughout the planet. Fortunately, the third stage unfolded mainly in the astral plane and on the lower levels of the mental plane - it was called the "cold war". In small countries, however, the war was still fought on the physical plane and was accompanied by abundant bloodshed, that is, it was definitely not "cold". Only after the forty-second year of the twentieth century did the dark forces finally begin to weaken, but more than forty years passed before a certain great disciple came to the levers of world power in 1985, under whom the end of the last stage of the war began and freedom and goodness began to spread again. will. But while the last flames of the world fire were dying out, new hotbeds of tension began to smolder in some places - mainly in those places where the god of money ruled. (Believers in him will sooner or later learn how vulnerable and fickle false gods are.)

Then, out of the ashes of the passing century, freedom first appeared in most of the world, and with it more Light. People interacted at such a pace and in so many ways that the forces of separation did not have time to interfere with them. Multinational corporations forced people to work together, and there was collaboration, at least on a professional level. More and more large state formations appeared, which coordinated their activities with others of the same kind (at first, mainly in the spheres of the economy and global security). Finally, it became clear that military force was losing its significance, and knowledge and information became more and more relevant. As a result, more and more forces began to focus on the study of the Earth, and then the near-Earth space. (Although the forces of darkness will continue to support the militarystrength at the expense of knowledge, art and culture.)

At the end of the millennium, many were waiting for some kind of global cataclysms to happen or even the end of the world. But nothing like that happened, and when the tension subsided, those same people for the first time felt the possibility of living in peace. It is hard to believe now that we humans have brought so much horror on ourselves and on each other. But the forces of darkness are finally "bound", and before us opens the opportunity to enter a new golden age. The Age of Pisces is being replaced by the Age of Aquarius, and group cooperation is replaced by individual fanaticism. Gotta seize the moment!

We are in for big changes.

As the twenty-first century dawned, amazing things began to happen. It has been noted that more and more organizations and even governments are led by enlightened leaders. For changing short-sighted, limited and short-sighted "leaders" came a new breed of people who saw a larger picture of the world and worked not for their own interests, but for the common good. After another couple of decades, the greatest blessing finally came: the World Teacher "reappeared" to help save the planet. Of course, many people still do not recognize the greatness of this Being, because it is in no way consistent with their prejudices. We are still slaves to our habits. Limited people supporting rigid belief systems, fiercely resist the wisdom that this great savior of the world demonstrates.

An enlightened leadership is being established throughout the planet. Colossal new energies are manifesting, both from higher planetary sources and from extraterrestrial realms, and we are finally entering the golden millennium. For all the time of the existence of mankind on the planet, such an era has not yet happened. Will it really be like that? Wait and see.

The Great Call

Around the middle of the twentieth century, an important spiritual tool was given to mankind. It is known as the Great Invocation. Its application and understanding is veryhelps the spiritual ascent of a person. First of all, it should be pointed out that we, people, are able to invoke Divine energies, which (although they are often ignored) are always available to us. With the advent of the Seventh Ray of ritual, rhythm and organization, the science of invocation - and this is precisely science - will increasingly enter into the consciousness of people, because correct invocation is exactly what an organized, rhythmic ritual is.

When prayer, meditation, hymn, etc., are used as an invocation and sincere efforts are made, by the law of resonance they evoke a response at higher levels. The more people use any call and the more often it is done, the more powerful and effective it becomes due to the cumulative effect. And the higher the level of spiritual consciousness in which the call is "packed", the greater its power. Engaging our higher spiritual consciousness in invoking high energies also ensures that these energies are used not for selfish purposes, but for the service of the whole world, to contribute to the enlightenment of our planet and all forms of life that exist on it. Here is the call:

From the point of Light that is in the Mind of God,

Let the Light flow into the minds of people.

Let Light descend upon the Earth.

From the point of Love in the Heart of God,

Let Love flow into the hearts of people.

May Christ return to earth.

From the Center where the Will of God is known,

Let the Purpose direct the small wills of people — The purpose, knowing which, the Teachers serve.

From the center of what we call the human race,

May the Plan of Love and Light will come true

And the door behind which evil will be sealed.

May Light, Love and Power be restored -

Plan on Earth.

As a person meditates on and uses the Great Invocation, it becomes increasingly clear to him that from this simple but very deep and powerful gift, humanity can draw many levels of meaning, aspects of perception (and practical results). I would like to present here what I call "scientific visualization" of the Great Invocation. In my opinion, the term "scientific" is justified by the fact that it corresponds to reality, and I will try to show this. And "visualization" in general is a fully conscious mental participation in the process that is to be realized. In other words, I will try to show how one can "see" the spiritual process at the levels at which we live and which, therefore, we can fully understand.

First Stanza:

From the point of Light that is in the Mind of God, Let

the Light stream into the minds of the people. Let Light

descend upon the Earth.

The "Point of Light that is in the mind of God" is higher, far higher than our highest understanding. This Light, the visible image of the Spirit, or higher consciousness, is born in what we can perceive as the mind (or mental aspect of the trinity) of God. From this point of purest intelligence, the Divine Light streams continuously into all the kingdoms of nature, including the Divine kingdoms, the human kingdom, the lower kingdoms, and those that are generally unknown to man. It is a consciousness that has always been infused and will always be infused into our minds. It is nothing but cosmic energy, the third aspect or Ray of the Divine Trinity. A huge force that brings humanity to an effective, reasonable level of great Life.
The end result of this is Enlightenment!

Light (or the consciousness of God) must descend from its levels and, if you like, fructify with itself all lives in all the kingdoms of our Earth. Over time, this leads to the growth and expansion of consciousness of all levels of being. If we imagine our Sun as a symbol (or lower correspondence) of the "Mind of God", and the light emitted by it as the personification of a higher mental plane, then we can see how these energies "stream", "descend to the Earth" and directly or indirectly penetrate into "minds of the people". On a physical level, we know that the Sun is the source of all life on the planet and through the action of sunlight (and alsosolar winds, sunspots, etc.), profound changes take place in all the kingdoms of nature.

Second Stanza:

From the point of Love in the Heart of God, Let Love flow

into the hearts of people. May Christ return to earth.

It is easy to imagine how the Light flows, but how to visualize Love?

I will focus on one of the reasons why this is not so easy to do. First of all, it must be emphasized that the first stanza is connected with the Third Ray of cosmic energy and, consequently, with the solarsystem that preceded ours. As a third ray solar system, it has given us at least the first idea of the Light. What we call "Divine Love" is still a new concept for us, as we are in the relatively early stages of our present solar system, which is the second solar system (in a series of three) and belongs to the Second Ray. It is in this solar system that Divine Love will be anchored on Earth. Although Divine Love is far from being fully materialized on the planes of our awareness, it seems to me that it is beginning to manifest itself in ways that are accessible to our perception. For example, I would suggest turning to color: passing through a prism, light forms colors, the seven spiritual colors. They may be one of the physicalmanifestations of love. Or take music: there are seven notes in an octave. To achieve harmony, one must be able to distinguish both sound and color, as well as know the measures and the right combinations. Studying harmonious proportions, we involuntarily plunge into the laws of geometry and mathematics, the golden section, etc.

All this leads to beauty, and beauty is the expression of Love in matter. Doesn't this mean that "the point of Love that is in the Heart of God" we, people, can we imagine as the center of the purest beauty, which, "flowing into our hearts", becomes compassion, altruism and everything that is best in a person? In the end, all these qualities, each in its own way, arose due to the ability to distinguish between the correct proportions and relationships. We know that the Divine Plan of Love ("Buddhic Plan") refers to the Second Ray of Love-Wisdom and with it such qualities expressing the right relationship as pure reason, intuition, mercy, a holistic worldview, compassion, altruism, etc. Therefore, I suggest that the beauty we perceive in art, music, architectural masterpieces, and other objects of the physical plane is the lowest reflection (which we can visualize) of the higher and more subtle qualities listed above. Visualizing "Love flowing into the hearts of people" (and into the heart of humanity), we can imagine beautiful colors and music - "the music of the spheres." (And the amazing beauty of nature.)

When we meet the word "Christ", we immediately remember the outstanding personality worshiped by Christians. But this great Being is better understood as the universal messenger of God who loves everyone regardless of religious beliefs. In the world he is known under a variety of names and titles. So: if we call for this great Being to descend further and further into matter, into the sphere where we inhabit - and this is exactly what is happening now - the "return of Christ to Earth" will certainly help us reach the hitherto unknown beauty of life.

Third Stanza:

From the center where God's will is known

Let the Purpose direct the small wills of people — *The purpose, knowing which, the Teachers serve.*

Who are the Teachers? These are developed beings who help the Savior of the World raise his consciousness. We call them Spiritual Mentors, Masters, Lords or Spiritual Hierarchs of our planet. Since this stanza refers to the energies of the First Ray, the key words here are "Will" and "Purpose." Let's talk about the goal first. As far as we can understand at our human level, the Divine Purpose is to raise and expand consciousness in all its manifestations. Or, in other words, to return the Universe to perfection through spiritual evolution.

Again, on a human level, this is accomplished by invoking the Third Ray Light energy, the Second Ray Love energy (verses one and two), and the First Ray Divine Will energy (verse three). But in the process of fulfilling the Divine Plan, constant purifications are necessary, because some entities resist enlightenment and need to be "remade" in order to get another chance. Part of the purification can be achieved through the destructive aspect of the First Ray. But here it should be emphasized: in fact, nothing can be destroyed - neither matter nor energy; everything is justturns into something else. Therefore, the First Ray does not destroy rather than transforms, releases or remakes.

Thus, the First Ray performs several functions: it energizes Light and Love; transforms what is needed, and also purifies, separating unliberated "atoms" for reworking. This can be visualized as follows: all impure (evil) is separated from the evolving life and washed into the center of the Earth for the fierypurification and transformation, and then again brought to the surface to repeat the process again. On the physical plane, we see how this happens in our body (the processes of digestion and excretion). Much attention is paid to Light and Love in the esoteric teachings, which cannot be said about the processes of purification and remaking. But this important and necessary activity is going on all the time, and we must participate in it consciously.

Fourth Stanza:

From the center of what we call the human race, May the

Plan of Love and Light come to pass,

And the door behind which evil.

Having invoked the illumination of the third ray, the compassionate wisdom of the second, and the focused power of the first, we return again to the throat "centre" of the planet: the human realm. Our job (dharma) is to fix "The Plan of Love and Light" so that its dynamic energies "fulfilled" first in our kingdom, and then in all the others (this is mentioned in the last stanza).

It is important to emphasize that everything in the universe is hierarchical (hierarchy means "sacred power"), and this is not a hierarchy of power, but rather of increasing responsibility. Each structural unit of the universe has the responsibility to help the representatives of the lower kingdoms. We humanity, together with the devas (angels), are those kingdoms best suited to support the animal, vegetable and mineral kingdoms. This is possible if you know the right ratios and proportions. Then we build our interaction with these kingdoms correctly and help the energies of Light, Love and Will to descend down to the less developed kingdoms and to the lower planes. And when all the kingdoms become enlightened, there will simply be no room for evil! By not participating in evil, we deprive it of its power, and this will help "seal" it so that it does not appear again. Therefore, we call for the sealing of the "door behind which evil" or unliberated, untransformed matter on the lower (gross) levels of all planes, which we, in fact, perceive as evil.

Fifth Stanza:

May Light, Love and Power be restored - Plan on Earth.

In the final stanza, we visualize "Light, Love and Power (Power)" emanating from the human (and higher) realms to "restore the (Divine) Plan on Earth." Can visualize myriads of points the lights of varying brightness that represent these realms, the energies of the third, second, and first rays already invoked, as well as Divine extra-planetary influences. All of this is in the right proportion and in the right relationship, interacting and spreading throughout the Earth system to help restore the Divine Plan of perfection from which humanity has temporarily deviated. Blessings to the readers of this book: In the name of Light, in the name of love, in the name of purpose we will try to fulfill its part of the One Cause. May it be so!

www.ingramcontent.com/pod-product-compliance
Lightning Source LLC
Chambersburg PA
CBHW052358220526
45465CB00003BB/1159